Reasonable Books = Reasonable Prices

Irina Kleyman

Exceptional for Home Schooling

Math Help Grade 2

AF882382

First Edition

Reasonable Books = Reasonable Prices

Irina Kleyman

Exceptional for Home Schooling

Math Help
Grade 2

I dedicate this book to my children and grandchildren.

Acknowledgments
Special thanks to Michael G LaFratta, Vladimir Kleyman,
who has assisted in bringing all 10 books to market.

Author's note: In the many years that I've taught mathematics, I have been informed by many of my students that my methods for teaching have helped them tremendously in learning math. I have written these books in order to continue to help others learn math effectively.

Math Help Grade 1
Math Help Grade 2
Math Help Grade 3
Math Help Grade 4
Math Help Grade 5
Math Help Grade 6
Math Help Grade 7
Math Help Grade 8
Math Help for College (Summary Grades 1 – 5)
Math Help for University (Basic, Algebra, Geometry)

ISBN: 978-1-7349699-1-7 Paperback

Printed in the United States of America.

Syllabus Math 1, 2, 3, 4

Grade 1 # 1 – 4 Grade 2 # 5 – 8 Grade 3 # 9-12 Grade 4 # 13 – 16

Syllabus includes topics, what students need to know.

Unknowns are numbers, which we need to find.

#	Grades	Topics. Advanced Problems (A-4), Pages (p)
	Chapters	Basic Problems (B), Chapter (C), Lesson (L)
1	1-1	One-Digit Numbers (B)
2	1-2	Two-Digit Numbers (B)
3	1-3	Three numbers (B)
4	1-4	Finding unknowns (A-1)
	Grade 1	Final Exam 1 B, C 4, L 12 # 1-20 odd, p 94
5	2-1	Addition or Subtraction Numbers (20–29) (B)
6	2-2	Addition or Subtraction Numbers (50–59) (B)
7	2-3	Addition or Subtraction Numbers (30–39) (B)
8	2-4	Finding unknowns (A-2)
	Grade 2	Final Exam 2 B, C 4, L 17, # 1 – 20 odd, p 106
9	3-1	Addition or Subtraction Numbers (60–69) (B)
10	3-2	Addition or Subtraction Numbers (90–99) (B)
11	3-3	Addition or Subtraction Numbers (70–79) (B)
12	3-4	Finding unknowns (A-3)
	Grade 3	Final Exam 3 B, C 4, L 19, # 1 – 20 odd, p 121
13	4-1	Three-Digit numbers (B)
14	4-2	Multiplication numbers (B)
15	4-3	Division numbers (B)
16	4-4	Finding unknowns (A-4)
	Grade 4	Final Exam 4 B, C 4, L 12, #1-10 all, p 133

Parents, Tutors, Students Middle School, High School, Colleges, University can help primary school students. Please read slowly math books to primary school students.

Grade 2

www.mathbooksbyirina.com

Chapter 1 Addition or Subtraction Numbers (20-29)

Gr[illegible] Chap[illegible]ntents
[illegible]dition[illegible]ract[illegible]bers (20–29)
[illegible]Lesson[illegible]
**[illegible]

Work with less[illegible] rules [illegible]ve the problems from contents, 3) Ch[illegible]rs in [illegible] 4) Work with the next lessons 2, 3, 4, 5[illegible]lve T[illegible]cale: A =10 points, B=9points, C=8[illegible]=7 p[illegible]-6 [illegible] If you solve 10 problems correctly, you g[illegible]) If h[illegible] yo[illegible]ve on to next chapter, 9) If your grad[illegible]han [illegible]rk [illegible]istakes, then solve Test 1B, 10) When you f[illegible] the [illegible]Exam in last chapter.

Lesson 1 One-d[illegible]ers.
Solve the probl[illegible]w 1A[illegible]8[illegible]B # 2, 4, 6, 8
Lesson 2 Two-d[illegible]ers.
Solve the probl[illegible]w 1A[illegible]8[illegible]B # 2, 4, 6, 8

Lesson 3 Three[illegible]
Solve the probl[illegible]w 1A[illegible]4[illegible]B # 1, 2, 3, 4
Lesson 4 Work[illegible]n: add[illegible]ubtract[illegible]rs in column.

Lesson 5 Add a[illegible]t N[illegible]
Solve the probl[illegible]w 1[illegible]8[illegible]B # 1, 2, 3, 4
Lesson 6 Numb[illegible] 21. [illegible] 20 (a, b, c, d)

Lesson 7 Numb[illegible] 23. [illegible] 20 (a, b, c, d)
Lesson 8 Numb[illegible] 25. [illegible] 20 (a, b, c, d)

Lesson 9 Numb[illegible] 27. [illegible] 20 (a, b, c, d)
Lesson 10 Num[illegible] 29. [illegible] # 20 (a, b, c, d)

Lesson 11 Test [illegible]inut[illegible]roble[illegible]
Solve Test 1 A [illegible] 7 (c[illegible]), # 10 (c, d)
Solve Test 1 B [illegible] 7 ([illegible]c, d[illegible]d), # 10 (c, d)
Answers Chapt[illegible]

Grade 2 Chapter 1 Lesson 1

Numbers from 0 to 9

Count forward from 1 to 9 **1, 2, 3, 4, 5, 6, 7, 8, 9**

Count backwards from 9 to 1 **9, 8, 7, 6, 5, 4, 3, 2, 1**

The smallest number is 0 **The biggest number is 9**

Equal means 5 = 5 **Number zero is less than one 0 < 1**

Less than means (<)

0 < 1	1 < 2	2 < 3	3 < 4	4 < 5
5 < 6	6 < 7	7 < 8	8 < 9	9 = 9

More than means (>)

9 > 8	8 > 7	7 > 6	6 > 5	5 > 4
4 > 3	3 > 2	2 > 1	1 > 0	0 = 0

Review 1 A

Add or Subtract numbers in row

	a	b	c	d	e
# 1	2 + 0	3 + 3	8 + 0	0 + 6	7 + 1
# 2	1 + 8	2 + 2	7 + 2	2 + 4	6 + 2
# 3	9 - 9	8 – 7	7 – 6	7 – 5	8 – 4
# 4	4 – 4	6 – 3	9 – 2	5 – 2	9 – 1

Add or Subtract numbers in column

	a	b	c	d	e
# 5	1 + 4	6 + 2	3 + 3	3 + 4	5 + 3
# 6	3 + 6	2 + 4	4 + 3	4 + 4	4 + 5
# 7	9 -8	7 - 7	6 - 6	6 - 5	7 - 4
# 8	9 -3	5 - 3	8 - 2	4 - 2	8 - 1

Grade 2 Chapter 1 Lesson 1

Addition numbers in row

The first operation with number is addition +

Property of addition

Two numbers can add in any order.
The sum wills the same. 2 + 3 = 5 3 + 2 = 5
The sum of a number and zero is the same as the number
5 + 0 = 5 0 + 5 = 5

Subtracting numbers in row

Second operation with numbers is subtraction –
Any number subtract zero result is the same number 5 – 0 = 5
Any number subtract the same number result is zero 5 – 5 = 0

Review 1 B

Add or Subtract numbers in row

	a	b	c	d	e
# 1	1 + 6	4 + 5	1 + 5	5 + 2	3 + 5
# 2	5 + 4	0 + 6	6 + 1	2 + 6	6 + 3
# 3	8 - 8	9 – 6	9 – 5	5 – 5	6 – 4
# 4	8 – 3	4 – 3	7 – 2	3 – 2	7 – 0

Add or Subtract numbers in column

	a	b	c	d	e
# 5	6 + 2	5 + 3	1 + 7	7 + 2	1 + 8
# 6	7 + 1	9 + 0	2 + 7	8 + 1	0 + 9
# 7	9 -7	8 - 6	8 - 5	9 - 4	5 - 4
# 8	7 -3	3 - 3	6 - 2	2 - 2	6 - 0

Grade 2 Chapter 1 Lesson 2

Two-digit numbers

One-digit numbers

One-digit numbers have one digit. This is numbers from 0 to 9
0, 1, 2, 3, 4, 5, 6, 7, 8, 9

Two-digit numbers

Two-digit numbers have two digits. Digits 0, 1, 2, 3, 4, 5, 6, 7, 8, 9
This is numbers from 10 to 99
Digit on the left are called tens. Digit on the right are called ones

Comparison numbers

One-digit numbers always less than two-digit numbers 9 < 11
Two-digit numbers always greater than one-digit numbers 19 > 8
Compare the tens and then compare the ones 12 < 18 18 > 12

Review 1 A

Add or Subtract numbers in row

	a	b	c	d	e
# 1	8 + 5	6 + 8	8 + 7	8 + 8	9 + 8
# 2	9 + 5	6 + 9	9 + 7	8 + 9	9 + 9
# 3	13 - 4	13 – 6	13 – 8	14 – 5	14 – 7
# 4	15 – 6	15 – 8	16 – 7	17 – 8	16 – 9

Add or Subtract numbers in column

	a	b	c	d	e
# 5	9 + 9	8 + 9	9 + 8	7 + 9	8 + 8
# 6	9 + 7	6 + 9	8 + 7	7 + 8	9 + 6
# 7	18 -9	17 - 9	16 - 8	15 - 6	15 - 8
# 8	14 -5	14 - 7	14 - 9	13 - 6	13 - 8

[illegible] Lesson 2

[illegible] write [illegible] two-digit numbers

Numbers from 10 [illegible] 10, [illegible], 14, 15, 16, 17, 18, 19

Numbers from 20 [illegible] 20, [illegible], 24, 25, 26, 27, 28, 29

Numbers from 30 [illegible] 30, [illegible], 34, 35, 36, 37, 38, 39

Numbers from 40 [illegible] 40, [illegible], 44, 45, 46, 47, 48, 49

Numbers from 50 [illegible] 50, [illegible], 54, 55, 56, 57, 58, 59

Numbers from 60 [illegible] 60, [illegible], 64, 65, 66, 67, 68, 69

Numbers from 70 [illegible] 70, [illegible], 74, 75, 76, 77, 78, 79

Numbers from 80 [illegible] 80, [illegible], 84, 85, 86, 87, 88, 89

Numbers from 90 [illegible] 90, [illegible], 94, 95, 96, 97, 98, 99

Review

Add [illegible] in row

	a	b	c	d	e
# 1	5 + 8	[illegible]	[illegible]	[illegible]	8 + 9
# 2	5 + 9	[illegible]	[illegible]	[illegible]	9 + 9
# 3	13 - 5	[illegible]	[illegible]	[illegible]	14 – 8
# 4	15 – 7	[illegible]	[illegible]	[illegible]	18 – 9

Add or subtract [illegible] in column

	a	b	c	d	e
# 5	9 + 5	[illegible]	[illegible]	[illegible]	5 + 9
# 6	8 + 5	[illegible]	[illegible]	[illegible]	4 + 9
# 7	17 -8	[illegible]	[illegible]	[illegible]	15 - 9
# 8	14 -6	[illegible]	13 [illegible]	[illegible]	13 - 9

Grade 2 Chapter 1 Lesson 3

Three Numbers

Addition three numbers in row

<u>**Three numbers can add in any order**</u>

9 + 4 + 3 = 13 + 3 = 16   **9 + 4 + 3 = 9 + 7 = 16**

Subtraction three numbers in row

<u>**To subtract numbers, work from left to right**</u>

10 – 1 – 4 = 9 – 4 = 5   **10 – 1 – 5 = 9 – 5 = 4**

Two operations with three numbers

<u>**To do two operations, work from left to right**</u>

5 + 6 – 9 = 11 – 9 = 2   **9 – 8 + 7 = 1 + 7 = 8**

5 + 8 – 9 = 13 – 9 = 4   **9 – 8 + 5 = 1 + 5 = 6**

Review 1 A

Add and Subtract numbers in row

	a	b	c	d	e
# 1	9 + 6 + 4	8 + 6 + 5	10 – 3 - 1	10 – 4 - 2	6 + 6 - 2
# 2	6 + 7 - 3	10 - 4 + 3	10 – 5 + 4	10 + 10	20 + 20

Add and Subtract numbers in column

	a	b	c	d	e
# 3	10 + 20	20 - 10	11 + 18	55 - 41	15 + 13 11
# 4	20 + 30	80 - 50	12 + 16	44 - 31	14 + 11 12

Grade 2 Chapter 1 Lesson 3

Adding or Subtraction numbers in column

Add or Subtract the ones and then add or subtract the tens

80	90	27	96	23
+ 10	- 10	+ 52	- 25	+ 32
90	80	79	71	21
				76

Review 1 B

Add and Subtract Numbers in row

	a	b	c	d	e
# 1	7 + 5 + 4	6 + 4 + 3	10 – 5 - 3	10 – 6 - 4	6 + 8 - 4
# 2	6 + 9 - 5	10 – 6 + 5	10 – 7 + 6	40 + 30	60 + 20

Add and Subtract Numbers in column

	a	b	c	d	e
# 3	30	70	73	74	42
	+ 40	- 30	+ 24	- 21	+ 20
					34
# 4	60	60	84	62	51
	+ 20	- 10	+12	- 11	+ 13
					23

Grade 2 Chapter 1 Lesson 4
Add or Subtract numbers in column
5 tens – 1 ten = 4 tens　　　10 ones = 1 ten

1) Number 50

1	4 10	1	4 10
a) 29	b) 5 0	c) 19	d) 50
+21	- 2 1	+31	- 19
50	2 9	50	31

2) Number 51

1	4 11	1	4 11
a) 29	b) 5 1	c) 19	d) 51
+22	- 2 2	+32	- 19
51	2 9	51	32

3) Number 52

1	4 12	1	4 12
a) 29	b) 5 2	c) 19	d) 52
+23	- 2 3	+33	- 19
52	2 9	52	33

4) Number 53

1	4 13	1	4 13
a) 29	b) 5 3	c) 19	d) 53
+24	- 2 4	+34	- 19
53	2 9	53	34

5) Numbers 54

1	4 14	1	4 14
1) 29	2) 5 4	3) 19	4) 54
+25	- 2 5	+ 35	- 19
54	2 9	54	35

G[illegible] Chap[illegible] [illegible]son 4

Add [illegible] in column

[illegible]s – 1 [illegible] tens [illegible] 0 ones = 1 ten

6) Number 55

1	4 1	1	4 15
[illegible] 29	[illegible] 5 5	[illegible] 19	d) 55
+26	-2 6	+36	- 19
55	2 9	55	36

7) Number 56

1	4 1	1	4 16
[illegible] 29	[illegible] 5 6	[illegible] 19	d) 56
+27	-2 7	+37	- 19
56	2 9	56	37

8) Number 57

1	4 1	1	4 17
[illegible] 29	[illegible] 5 7	[illegible] 19	d) 57
+28	-2 8	+38	- 19
57	2 9	57	38

9) Number 58

1	4 1	1	4 18
[illegible] 29	[illegible] 5 8	[illegible] 19	d) 58
+29	-2 9	+39	- 19
58	2 9	58	39

10) Number 59

[illegible]	[illegible]	[illegible] 19	d) 59
+[illegible]	-30	+40	-19
[illegible]	[illegible]	59	40

Grade 2 Chapter 1 Lesson 5
Addition and Subtraction Numbers
Review 1 A
Add or Subtract numbers in row

	a	b	c	d	e
# 1	8 + 5	6 + 8	8 + 7	8 + 8	9 + 8
# 2	9 + 5	6 + 9	9 + 7	8 + 9	9 + 9
# 3	13 - 4	13 – 6	13 – 8	14 – 5	14 – 7
# 4	15 – 6	15 – 8	16 – 7	17 – 8	16 – 9

Add or Subtract numbers in column

	a	b	c	d	e
# 5	9 + 9	8 + 9	9 + 8	7 + 9	8 + 8
# 6	9 + 7	6 + 9	8 + 7	7 + 8	9 + 6
# 7	18 -9	17 - 9	16 - 8	15 - 6	15 - 8
# 8	14 -5	14 - 7	14 - 9	13 - 6	13 - 8

Review 1 B
Add and Subtract numbers in row

	a	b	c	d	e
# 1	8 + 6 + 4	9 + 4 + 5	10 – 4 - 1	10 – 5 - 1	7 + 6 - 2
# 2	8 + 7 - 3	12 - 4 + 3	11 – 5 + 4	10 + 20	20 + 20

Add and Subtract numbers in column

	a	b	c	d	e
# 3	10 + 20	80 - 70	12 + 16	55 - 31	14 + 12 11
# 4	20 + 30	60 - 40	13 + 16	54 - 31	15 + 12 12

Grade 2 Chapter 1 Lesson 6

Number 20

Remember

1+ 9 = 2 + 8 = 3 + 7 = 4 + 6 = 5 + 5 = 10

Add or Subtract numbers in row

6 + 14 = 14 + 6 =10+4+6=10+10= 20

20 – 14 = 20 – 10 – 4 = 10 – 4 = 6

Review 1 A

	a	b	c	d
# 1	20 + 0	20 - 0	10 + 10	20 - 10
# 2	19 + 1	20 – 1	9 + 11	20 – 11
# 3	18 + 2	20 - 2	8 + 12	20 - 12
# 4	17 + 3	20 - 3	7 + 13	20 - 13
# 5	16 + 4	20 – 4	6 + 14	20 – 14
# 6	15 + 5	20 - 5	5 + 15	20 - 15
# 7	14 + 6	20 – 6	4 + 16	20 – 16
# 8	13 + 7	20 - 7	3 + 17	20 - 17
# 9	12 + 8	20 – 8	2 + 18	20 – 18
# 10	11 + 9	20 – 9	1 + 19	20 - 19

Number 21

Add or Subtract numbers in row

7 + 14 = 14 + 7 = 10 + 4 + 7 =10 + 11 = 21

21 – 14 = 21 – 11 – 3 = 10 – 3 = 7

Review 1 B

	a	b	c	d
# 11	21 + 0	21 - 0	11 + 10	21 - 10
# 12	20 + 1	21 – 1	10 + 11	21 – 11
# 13	19 + 2	21 - 2	9 + 12	21 - 12
# 14	18 + 3	21 – 3	8 + 13	21 – 13
# 15	17 + 4	21 - 4	7 + 14	21 – 14
# 16	16 + 5	21 – 5	6 + 15	21 – 15
# 17	15 + 6	21 - 6	5 + 16	21 - 16
# 18	14 + 7	21 – 7	4 + 17	21 – 17
# 19	13 + 8	21 - 8	3 + 18	21 - 18
# 20	12 + 9	21 – 9	2 + 19	21 – 19

Grade 2 Chapter 1 Lesson 7

Number 22

Remember

1+ 9 = 2 + 8 = 3 + 7 = 4 + 6 = 5 + 5 = 10

Add or Subtract numbers in row

4 + 18 = 18 + 4 = 10 + 8 + 4 = 10 + 12 = 22

22 – 14 = 22 – 12 – 2 = 10 – 2 = 8

Review 1 A

	a	b	c	d
# 1	22 + 0	22 - 0	12 + 10	22 - 10
# 2	21 + 1	22 –1	11 + 11	22 – 11
# 3	20 + 2	22 - 2	10 + 12	22 - 12
# 4	19 + 3	22 –3	9 + 13	22 – 13
# 5	18 + 4	22 - 4	8 + 14	22 – 14
# 6	17 + 5	22 –5	7 + 15	22 – 15
# 7	16 + 6	22 - 6	6 + 16	22 - 16
# 8	15 + 7	22 –7	5 + 17	22 – 17
# 9	14 + 8	22 - 8	4 + 18	22 - 18
# 10	13 + 9	22 –9	3 + 19	22 – 19

Number 23

9 + 14 = 14 + 9 = 10 + 4 + 9=10+13 = 23

23 – 14 = 23 – 13 – 1 = 10 – 1 = 9

Review 1 B

	a	b	c	d
# 11	23 + 0	23 - 0	13 + 10	23 - 10
# 12	22 + 1	23 – 1	12 + 11	23 – 11
# 13	21 + 2	23 - 2	11 + 12	23 - 12
# 14	20 + 3	23 – 3	10 + 13	23 – 13
# 15	19 + 4	23 - 4	9 + 14	23 – 14
# 16	18 + 5	23 – 5	8 + 15	23 –15
# 17	17 + 6	23 - 6	7 + 16	23 - 16
# 18	16 + 7	23 – 7	6 + 17	23 –17
# 19	15 + 8	23 - 8	5 + 18	23 - 18
# 20	14 + 9	23 – 9	4 + 19	23 -19

Gra[illegible] 8

[illegible]

1 + 9 = [illegible] 5 + 5 = 10

[illegible]

10 + [illegible] + 4 = 24

24 – [illegible] – 4 = 10

[illegible]

Ad[illegible] s in row

	a	b	c	d
# 1	24 [illegible]	[illegible] - 0	[illegible]	[illegible] - 10
# 2	23 [illegible]	[illegible] – 1	[illegible]	[illegible] – 11
# 3	22 [illegible]	[illegible] - 2	[illegible]	[illegible] - 12
# 4	21 [illegible]	[illegible] – 3	[illegible]	[illegible] – 13
# 5	20 [illegible]	[illegible] - 4	[illegible]	[illegible] – 14
# 6	19 [illegible]	[illegible] – 5	[illegible]	[illegible] – 15
# 7	18 [illegible]	[illegible] - 6	[illegible]	[illegible] - 16
# 8	17 [illegible]	[illegible] – 7	[illegible]	[illegible] – 17
# 9	16 [illegible]	[illegible] - 8	[illegible]	[illegible] - 18
# 10	15 [illegible]	[illegible] – 9	[illegible]	[illegible] – 19

[illegible]

[illegible] 10 + [illegible] 1 + 4 = 20 + 5 = 25

25 – [illegible] – 4 = 11

[illegible]

Ad[illegible] s in row

	a	b	c	d
# 11	25 [illegible]	[illegible] - 0	[illegible]	[illegible] - 10
# 12	24 [illegible]	[illegible] – 1	[illegible]	[illegible] –11
# 13	23 [illegible]	[illegible] - 2	[illegible]	[illegible] - 12
# 14	22 [illegible]	[illegible] – 3	[illegible]	[illegible] –13
# 15	21 [illegible]	[illegible] - 4	[illegible]	[illegible] – 14
# 16	20 [illegible]	[illegible] – 5	[illegible]	[illegible] –15
# 17	19 [illegible]	[illegible] - 6	[illegible]	[illegible] - 16
# 18	18 [illegible]	[illegible] – 7	[illegible]	[illegible] –17
# 19	17 [illegible]	[illegible] - 8	[illegible]	[illegible] - 18
# 20	16 [illegible]	[illegible] – 9	[illegible]	[illegible] –19

Grade 2 Chapter 1 Lesson 9

Remember

1+ 9 = 2 + 8 = 3 + 7 = 4 + 6 = 5 + 5 = 10

Numbers 26

12 + 14 = 10 + 2 + 10 + 4 = 10 + 10 + 2 + 4 = 20 + 6 = 26

26 - 14 = 26 – 10 – 4 = 16 – 4 = 12

Review 1 A

Add or Subtract numbers in row

	a	b	c	d
# 1	26 + 0	26 - 0	16 + 10	26 - 10
# 2	25 + 1	26 – 1	15 + 11	26 – 11
# 3	24 + 2	26 - 2	14 + 12	26 - 12
# 4	23 + 3	26 – 3	13 + 13	26 – 13
# 5	22 + 4	26 - 4	12 + 14	26 – 14
# 6	21 + 5	26 – 5	11 + 15	26 – 15
# 7	20 + 6	26 - 6	10 + 16	26 - 16
# 8	19 + 7	26 – 7	9 + 17	26 – 17
# 9	18 + 8	26 - 8	8 + 18	26 - 18
# 10	17 +9	26 – 9	7 + 19	26 – 19

Number 27

13 + 14 = 10 + 3 + 10 + 4 = 10 + 10 + 3 + 4 = 20 + 7 = 27

27 – 14 = 27 – 10 – 4 = 17 – 4 = 13

Review 1 B

Add or Subtract numbers in row

	a	b	c	d
# 11	27 + 0	27 - 0	17 + 10	27 - 10
# 12	26 + 1	27 – 1	16 + 11	27 –11
# 13	25 + 2	27 - 2	15 + 12	27 - 12
# 14	24 + 3	27 – 3	14 + 13	27 –13
# 15	23 + 4	27 - 4	13 + 14	27 – 14
# 16	22 + 5	27 – 5	12 + 15	27 –15
# 17	21 + 6	27 - 6	11 + 16	27 - 16
# 18	20 + 7	27 – 7	10 + 17	27 –17
# 19	19 + 8	27 - 8	9 + 18	27 - 18
# 20	18 + 9	27 – 9	8 + 19	27 - 19

Grade 2 Chapter 1 Lesson 10

Remember

1+ 9 = 2 + 8 = 3 + 7 = 4 + 6 = 5 + 5 = 10

Number 28

14 + 14 = 10 + 4 + 10 + 4 = 10 + 10 + 4 + 4 = 20 + 8 = 28

28 – 14 = 28 – 10 – 4 = 18 – 4 = 14

Review 1 A

Add or Subtract numbers in row

	a	b	c	d
# 1	28 + 0	28 - 0	18 + 10	28 - 10
# 2	27 + 1	28 – 1	17 + 11	28 – 11
# 3	26 + 2	28 - 2	16 + 12	28 - 12
# 4	25 + 3	28 – 3	15 + 13	28 – 13
# 5	24 + 4	28 - 4	14 + 14	28 – 14
# 6	23 + 5	28 – 5	13 + 15	28 –15
# 7	22 + 6	28 - 6	12 + 16	28 – 16
# 8	21 + 7	28 – 7	11 + 17	28 –17
# 9	20 + 8	28 - 8	10 + 18	28 – 18
# 10	19 + 9	28 – 9	9 + 19	28 –19

Number 29

15 + 14 = 10 + 5 + 10 + 4 = 10 + 10 + 5 + 4 = 20 + 9 = 29

29 – 14 = 29 – 10 – 4 = 19 – 4 = 15

Review 1 B

Add or Subtract numbers in row

	a	b	c	d
# 11	29 + 0	29 - 0	19 + 10	29 - 10
# 12	28 + 1	29 – 1	18 + 11	29 – 11
# 13	27 + 2	29 - 2	17 + 12	29 - 12
# 14	26 + 3	29 – 3	16 + 13	29 – 13
# 15	25 + 4	29 - 4	15 + 14	29 - 14
# 16	24 + 5	29 – 5	14 + 15	29 – 15
# 17	23 + 6	29 - 6	13 + 16	29 - 16
# 18	22 + 7	29 – 7	12 + 17	29 – 17
# 19	21 + 8	29 - 8	11 + 18	29 - 18
# 20	20 + 9	29 – 9	10 + 19	29 – 19

Grade 2 Chapter 1 Lesson 11

Addition or Subtraction Numbers in row

Numbers 20 to 29

Test 1 A (50 minutes)

Add or Subtract Numbers in row

	a	b	c	d
# 1	20 + 0	20 - 0	10 + 10	20 - 10
# 2	19 + 1	20 – 1	9 + 11	20 – 11
# 3	19 + 2	21 - 2	9 + 12	21 - 12
# 4	18 + 3	21 – 3	8 + 13	21 – 13
# 5	18 + 4	22 - 4	8 + 14	22 - 14
# 6	17 + 5	22 –5	7 + 15	22 – 15
# 7	17 + 6	23 - 6	7 + 16	23 - 16
# 8	16 + 7	23 – 7	6 + 17	23 –17
# 9	16 + 8	24 - 8	6 + 18	24 - 18
# 10	15 + 9	24 – 9	5 + 19	24 – 19

Gr[illegible] hap[illegible] on 11

[illegible]dition[illegible] rac[illegible] bers in row

[illegible] umbers [illegible]

[illegible] B (5[illegible] s)

Ad[illegible] act[illegible] s in row

	a	b	c	d
# 1	17 – [illegible]	[illegible]5 - [illegible]	[illegible] – 18	[illegible] - 18
# 2	16 – [illegible]	[illegible]5 – [illegible]	[illegible] – 19	[illegible] –19
# 3	26 – [illegible]	[illegible]6 - [illegible]	[illegible] + [illegible]	[illegible] - 10
# 4	25 – [illegible]	[illegible]6 – [illegible]	[illegible] + [illegible]	[illegible] – 11
# 5	25 – [illegible]	[illegible]7 - [illegible]	[illegible] + [illegible]	[illegible]7 - 12
# 6	24 – [illegible]	[illegible]7 – [illegible]	[illegible] + [illegible]	[illegible]7 –13
# 7	26 – [illegible]	[illegible]8 - [illegible]	[illegible] + [illegible]	[illegible]8 - 12
# 8	25 – [illegible]	[illegible]8 – [illegible]	[illegible] + [illegible]	[illegible]8 – 13
# 9	25 – [illegible]	[illegible]9 - [illegible]	[illegible] + [illegible]	[illegible]9 - 14
# 10	24 – [illegible]	[illegible]9 – [illegible]	[illegible] + [illegible]	[illegible]9 – 15

Grade 2 Answers Chapter 1

Lesson 1

Review 1 A

	a	b	c	d	e		a	b	c	d	e
# 1	2	6	8	6	8	# 5	5	8	6	7	8
# 2	9	4	9	6	8	# 6	9	6	7	8	9
# 3	0	1	1	2	4	# 7	1	0	0	1	3
# 4	0	3	7	3	8	# 8	6	2	6	2	7

Review 1 B

	a	b	c	d	e		a	b	c	d	e
# 1	7	9	6	7	8	# 5	8	8	8	9	9
# 2	9	6	7	8	9	# 6	8	9	9	9	9
# 3	0	3	4	0	2	# 7	2	2	3	5	1
# 4	5	1	5	1	7	# 8	4	0	4	0	6

Lesson 2

Review 1 A

	a	b	c	d	e		a	b	c	d	e
# 1	13	14	15	16	17	# 5	18	17	17	16	16
# 2	14	15	16	17	18	# 6	16	15	15	15	15
# 3	9	7	5	9	7	# 7	9	8	8	9	7
# 4	9	7	9	9	7	# 8	9	7	5	7	5

Review 1 B

	a	b	c	d	e		a	b	c	d	e
# 1	13	14	15	16	17	# 5	14	14	14	14	14
# 2	14	15	16	17	18	# 6	13	13	13	13	13
# 3	8	6	4	8	6	# 7	9	9	7	8	6
# 4	8	6	8	8	9	# 8	8	6	8	6	4

Lesson 3

Review 1 A

	a	b	c	d	e		a	b	c	d	e
# 1	19	19	6	4	10	# 3	30	10	29	14	39
# 2	10	9	9	20	40	# 4	50	30	28	13	37

Review 1 B

	a	b	c	d	e		a	b	c	d	e
# 1	16	13	2	0	10	# 3	70	40	97	53	96
# 2	10	9	9	70	80	# 4	80	50	96	51	87

Lesson 5
Review 1 A

	a	b	c	d	e		a	b	c	d	e
# 1	13	14	15	16	17	# 5	18	17	17	16	16
# 2	14	15	16	17	18	# 6	16	15	15	15	15
# 3	9	7	5	9	7	# 7	9	8	8	9	7
# 4	9	7	9	9	7	# 8	9	7	5	7	5

Review 1 B

	a	b	c	d	e		a	b	c	d	e
# 1	18	18	5	4	11	# 3	30	10	28	24	37
# 2	12	11	10	30	40	# 4	50	20	29	23	39

Lesson 6

	a	b	c	d		a	b	c	d
# 1	20	20	20	10	# 11	21	21	21	11
# 2	20	19	20	9	# 12	21	20	21	10
# 3	20	18	20	8	# 13	21	19	21	9
# 4	20	17	20	7	# 14	21	18	21	8
# 5	20	16	20	6	# 15	21	17	21	7
# 6	20	15	20	5	# 16	21	16	21	6
# 7	20	14	20	4	# 17	21	15	21	5
# 8	20	13	20	3	# 18	21	14	21	4
# 9	20	12	20	2	# 19	21	13	21	3
# 10	20	11	20	1	# 20	21	12	21	2

Grade 2 Answers Chapter 1

Lesson 7

	a	b	c	d		a	b	c	d
# 1	22	22	22	12	# 11	23	23	23	13
# 2	22	21	22	11	# 12	23	22	23	12
# 3	22	20	22	10	# 13	23	21	23	11
# 4	22	19	22	9	# 14	23	20	23	10
# 5	22	18	22	8	# 15	23	19	23	9
# 6	22	17	22	7	# 16	23	18	23	8
# 7	22	16	22	6	# 17	23	17	23	7
# 8	22	15	22	5	# 18	23	16	23	6
# 9	22	14	22	4	# 19	23	15	23	5
# 10	22	13	22	3	# 20	23	14	23	4

Lesson 8

	a	b	c	d		a	b	c	d
# 1	24	24	24	14	# 11	25	25	25	15
# 2	24	23	24	13	# 12	25	24	25	14
# 3	24	22	24	12	# 13	25	23	25	13
# 4	24	21	24	11	# 14	25	22	25	12
# 5	24	20	24	10	# 15	25	21	25	11
# 6	24	19	24	9	# 16	25	20	25	10
# 7	24	18	24	8	# 17	25	19	25	9
# 8	24	17	24	7	# 18	25	18	25	8
# 9	24	16	24	6	# 19	25	17	25	7
# 10	24	15	24	5	# 20	25	16	25	6

Lesson 9

	a	b	c	d		a	b	c	d
# 1	26	26	26	16	# 11	27	27	27	17
# 2	26	25	26	15	# 12	27	26	27	16
# 3	26	24	26	14	# 13	27	25	27	15
# 4	26	23	26	13	# 14	27	24	27	14
# 5	26	22	26	12	# 15	27	23	27	13
# 6	26	21	26	11	# 16	27	22	27	12
# 7	26	20	26	10	# 17	27	21	27	11
# 8	26	19	26	9	# 18	27	20	27	10
# 9	26	18	26	8	# 19	27	19	27	9
# 10	26	17	26	7	# 20	27	18	27	8

Lesson 10

	a	b		d
# 1	28	28	[illegible]	19
# 2	28	27	[illegible]	18
# 3	28	26	[illegible]	17
# 4	28	25	[illegible]	16
# 5	28	24	[illegible]	15
# 6	28	23	[illegible]	14
# 7	28	22	[illegible]	13
# 8	28	21	[illegible]	12
# 9	28	20	[illegible]	11
# 10	28	19	[illegible]	10

Lesson 11

Test 1 A

	a	b		d
# 1	20	20	[illegible]	7
# 2	20	19	[illegible]	7
# 3	21	19	[illegible]	6
# 4	21	18	[illegible]	6
# 5	22	18	[illegible]	5

Test 1 B

	a	b		d
# 1	25	17	[illegible]	14
# 2	25	16	[illegible]	16
# 3	26	26	[illegible]	15
# 4	26	25	[illegible]	15
# 5	27	25	[illegible]	14

Grade 2

Chapter 2 Addition or Subtraction Numbers (50-59)

Grade 2 Chapter 2 Contents
Addition or Subtraction Numbers (50 – 59)
Lesson 1 - 11
Basic Problems

Work with lesson 1, read rules and examples, 2) Solve the problems from contents, 3) Check answers in the end of chapter 1, 4) Work with the next lessons 2, 3, 4, 5, …, 5) Solve Test 1 A, 6) Grading Scale: A =10 points, B=9points, C=8 points, D=7 points, F=0-6 points, 7) If you solve 10 problems correctly, you get an A, 8) If have an A, you can move on to next chapter, 9) If your grade is lower than an A, work with the mistakes, then solve Test 1B, 10) When you finish all of the chapters, solve Final Exam in last chapter.

Grade 2 Chapter 2 Lesson 1
Number 50
Add or Subtract numbers in column

5 tens – 1 ten = 4 tens **10 ones = 1 ten**

a)	b)	c)	d)
1	4 10	1	4 10
29	5 0	19	50
+21	- 2 1	+31	- 19
50	2 9	50	31

Add or Subtract

	a	b	c	d	e
# 1	50 + 0	50 - 0	30 + 20	50 - 20	10 + 40
# 2	49 + 1	50 - 1	29 +21	50 - 21	9 + 41
# 3	48 + 2	50 - 2	28 + 22	50 - 22	8 + 42
# 4	47 + 3	50 - 3	27 + 23	50 - 23	7 + 43
# 5	46 + 4	50 - 4	26 + 24	50 - 24	6 + 44
# 6	45 + 5	50 - 5	25 + 25	50 - 25	5 + 45
# 7	44 + 6	50 - 6	24 + 26	50 - 26	4 + 46
# 8	43 + 7	50 - 7	23 + 27	50 - 27	3 + 47
# 9	42 + 8	50 - 8	22 + 28	50 - 28	2 + 48
# 10	41 + 9	50 - 9	21 + 29	50 - 29	1 + 49

G[illegible] Chap[illegible] [illegible]son 1

	a	b	c	d
# 11	40 +10	[illegible]	20 + 30	[illegible]
# 12	39 +11	[illegible]	19 + 31	[illegible]
# 13	38 +12	[illegible]	18 + 32	[illegible]
# 14	37 +13	[illegible]	17 + 33	[illegible]
# 15	36 + 14	[illegible]	16 + 34	[illegible]
# 16	35 +15	[illegible]	15 + 35	[illegible]
# 17	34 +16	[illegible]	14 + 36	[illegible]
# 18	33 +17	[illegible]	13 + 37	[illegible]
# 19	32 +18	[illegible]	12 +38	[illegible]
# 20	31 +19	[illegible]	11 +39	[illegible]

Grade 2 Chapter 2 Lesson 2
Number 51
Add or Subtract numbers in column

5 tens – 1 ten = 4 tens   1 ten = 10 ones

a)	b)	c)	d)
1	4 11	1	4 11
29	5 1	19	51
+22	- 2 2	+32	- 19
51	2 9	51	32

Add or Subtract

	a	b	c	d	e
# 1	51 +0	51 - 0	31 +20	51 -20	11 +40
# 2	50 +1	51 -1	30 +21	51 -21	10 + 41
# 3	49 + 2	51 - 2	29 +22	51 -22	9 +42
# 4	48 + 3	51 - 3	28 +23	51 - 23	8 +43
# 5	47 + 4	51 -4	27 +24	51 -24	7 +44
# 6	46 + 5	51 -5	26 +25	51 -25	6 +45
# 7	45 + 6	51 - 6	25 + 26	51 - 26	5 + 46
# 8	44 + 7	51 - 7	24 +27	51 -27	4 + 47
# 9	43 + 8	51 - 8	23 + 28	51 -28	3 +48
# 10	42 + 9	51 - 9	22 +29	51 -29	2 +49

Grade 2 Chapter 2 Lesson 2
Number 51
Add or Subtract

	a	b	c	d	e
# 11	41 +10	51 -10	21 +30	51 -30	51 - 40
# 12	40 + 11	51 -11	20 +31	51 -31	51 - 41
# 13	39 + 12	51 -12	19 +32	51 -32	51 -42
# 14	38 + 13	51 -13	18 +33	51 -33	51 -43
# 15	37 + 14	51 - 14	17 +34	51 -34	51 -44
# 16	36 + 15	51 -15	16 +35	51 -35	51 -45
# 17	35 + 16	51 - 16	15 +36	51 -36	51 -46
# 18	34 + 17	51 -17	14 +37	51 -37	51 -47
# 19	33 + 18	51 - 18	13 +38	51 - 38	51 - 48
# 20	32 + 19	51 - 19	12 +39	51 -39	51 -49

Grade 2 Chapter 2 Lesson 3
Number 52
5 tens – 1 ten = 4 tens   1 ten = 10 ones

a)	b)	c)	d)
1	4 12	1	4 12
29	5 2	19	52
+23	- 2 3	+33	- 19
52	2 9	52	33

Add or Subtract numbers in column

	a	b	c	d	e
# 1	52 +0	52 -0	32 +20	52 -20	12 +40
# 2	51 +1	52 -1	31 +21	52 -21	11 +41
# 3	50 +2	52 -2	30 +22	52 - 22	10 +42
# 4	49 +3	52 -3	29 +23	52 -23	9 +43
# 5	48 +4	52 -4	28 +24	52 -24	8 +44
# 6	47 +5	52 -5	27 +25	52 -25	7 +45
# 7	46 +6	52 -6	26 +26	52 -26	6 +46
# 8	45 +7	52 - 7	25 +27	52 -27	5 +47
# 9	44 +8	52 - 8	24 +28	52 -28	4 +48
# 10	43 +9	52 - 9	23 +29	52 -29	3 +49

[illegible] son 3

	a
# 11	42 +10
# 12	41 +11
# 13	40 +12
# 14	39 +13
# 15	38 +14
# 16	37 +15
# 17	36 +16
# 18	35 +17
# 19	34 +18
# 20	33 +19

Grade 2 Chapter 2 Lesson 4
Number 53
5 tens – 1 ten = 4 tens   1 ten = 10 ones

a)	b)	c)	d)
1	4 13	1	4 13
29	5 3	19	53
+24	- 2 4	+34	- 19
53	2 9	53	34

Add or Subtract numbers in column

	a	b	c	d	e
# 1	53	53	33	53	13
	+ 0	- 0	+20	-20	+40
# 2	52	53	32	53	12
	+ 1	-1	+21	-21	+41
# 3	51	53	31	53	11
	+ 2	- 2	+22	-22	+42
# 4	50	53	30	53	10
	+ 3	- 3	+23	-23	+43
# 5	49	53	29	53	9
	+ 4	- 4	+24	-24	+44
# 6	48	53	28	53	8
	+ 5	- 5	+25	-25	+45
# 7	47	53	27	53	7
	+ 6	- 6	+26	-26	+46
# 8	46	53	26	53	6
	+ 7	- 7	+27	-27	+47
# 9	45	53	25	53	5
	+ 8	- 8	+28	-28	+48
# 10	44	53	24	53	4
	+9	- 9	+29	-29	+49

Grade 2 Chapter 2 Lesson 4
Number 53
Add or Subtract

	a	b	c	d	e
# 11	43	53	23	53	53
	+ 10	-10	+30	-30	-40
# 12	42	53	22	53	53
	+ 11	-11	+31	-31	- 41
# 13	41	53	21	53	53
	+ 12	-12	+32	-32	-42
# 14	40	53	20	53	53
	+ 13	- 13	+33	-33	-43
# 15	39	53	19	53	53
	+ 14	-14	+34	-34	-44
# 16	38	53	18	53	53
	+ 15	- 15	+35	-35	- 45
# 17	37	53	17	53	53
	+ 16	-16	+36	-36	-46
# 18	36	53	16	53	53
	+ 17	- 17	+37	-37	-47
# 19	35	53	15	53	53
	+ 18	-18	+38	-38	-48
# 20	34	53	14	53	53
	+ 19	-19	+39	-39	-49

Grade 2 Chapter 2 Lesson 5
Number 54
Add or Subtract numbers in column
5 tens – 1 ten = 4 tens   1 ten = 10 ones

a)	b)	c)	d)
1	4 14	1	4 14
29	5 4	19	54
+25	- 2 5	+35	- 19
54	2 9	54	35

Add or Subtract numbers in column

	a	b	c	d	e
# 1	54 +0	54 -0	34 +20	54 -20	14 +40
# 2	53 +1	54 -1	33 +21	54 -21	13 +41
# 3	52 +2	54 -2	32 +22	54 -22	12 +42
# 4	51 +3	54 -3	31 +23	54 -23	11 +43
# 5	50 +4	54 -4	30 +24	54 -24	10 +44
# 6	49 +5	54 -5	29 +25	54 -25	9 +45
# 7	48 +6	54 -6	28 +26	54 -26	8 +46
# 8	47 +7	54 -7	27 +27	54 -27	7 +47
# 9	46 +8	54 -8	26 +28	54 -28	6 +48
# 10	45 +9	54 -9	25 +29	54 -29	7 +47

Gr[illegible] Chap[illegible] [illegible]son 5

[illegible]

	a				e
# 11	44 +10	[illegible]	[illegible]	[illegible]	54 +40
# 12	43 +11	[illegible]	[illegible]	[illegible]	54 +41
# 13	42 +12	[illegible]	[illegible]	[illegible]	54 +42
# 14	41 +13	[illegible]	[illegible]	[illegible]	54 +43
# 15	40 +14	[illegible]	[illegible]	[illegible]	54 +44
# 16	39 +15	[illegible]	[illegible]	[illegible]	54 +45
# 17	38 +16	[illegible]	[illegible]	[illegible]	54 +46
# 18	37 +17	[illegible]	[illegible]	[illegible]	54 +47
# 19	36 +18	[illegible]	[illegible]	[illegible]	54 +48
# 20	35 +19	[illegible]	[illegible]	[illegible]	54 +49

Grade 2 Chapter 2 Lesson 6
Number 55
5 tens – 1 ten = 4 tens   1 ten = 10 ones

a)	b)	c)	d)
1	4 15	1	4 15
29	5 5	19	55
+26	- 2 6	+36	- 19
55	2 9	55	36

Add or Subtract numbers in column

	a	b	c	d	e
# 1	55 +0	55 -0	35 +20	55 -20	15 +40
# 2	54 +1	55 -1	34 +21	55 -21	14 +41
# 3	53 +2	55 -2	33 +22	55 -22	13 +42
# 4	52 +3	55 -3	32 +23	55 -23	12 +43
# 5	51 +4	55 -4	31 +24	55 -24	11 +44
# 6	50 +5	55 -5	30 + 25	55 -25	10 +45
# 7	49 +6	55 -6	29 +26	55 -26	9 +46
# 8	48 +7	55 -7	28 + 27	55 - 27	8 +47
# 9	47 +8	55 -8	27 +28	55 -28	7 +48
# 10	46 +9	55 -9	26 +29	55 -29	6 +49

Grade 2 Chapter 2 Lesson 6
Number 55
Add or Subtract

	a	b	c	d	e
# 11	45 +10	55 -10	25 +30	55 -30	55 -40
# 12	44 + 11	55 -11	24 + 31	55 -31	55 -41
# 13	43 +12	55 -12	23 +32	55 -32	55 -42
# 14	42 +13	55 -13	22 +33	55 -33	55 -43
# 15	41 +14	55 -14	21 +34	55 -34	55 -44
# 16	40 +15	55 -15	20 +35	55 -35	55 -45
# 17	39 +16	55 -16	19 +36	55 -36	55 -46
# 18	38 +17	55 -17	18 +37	55 -37	55 -47
# 19	37 +18	55 -18	17 +38	55 -38	55 -48
# 20	36 +19	55 -19	16 +39	55 -39	55 -49

Grade 2 Chapter 2 Lesson 7
Number 56
Add or Subtract numbers in column
5 tens – 1 ten = 4 tens   1 ten = 10 ones

a)	b)	c)	d)
1	4 16	1	4 16
29	5 6	19	56
+27	- 2 7	+37	- 19
56	2 9	56	37

Add or Subtract numbers in column

	a	b	c	d	e
# 1	56 +0	56 -0	36 +20	56 -20	16 +40
# 2	55 +1	56 -1	35 +21	56 -21	15 +41
# 3	54 +2	56 -2	34 +22	56 -22	14 +42
# 4	53 +3	56 -3	33 +23	56 -23	13 +43
# 5	52 +4	56 -4	32 +24	56 -24	12 +44
# 6	51 +5	56 -5	31 + 25	56 -25	11 +45
# 7	50 +6	56 -6	30 +26	56 -26	10 +46
# 8	49 +7	56 -7	29 +27	56 -27	9 +47
# 9	48 +8	56 -8	28 +28	56 -28	8 +48
# 10	47 +9	56 -9	27 +29	56 -29	7 + 49

[illegible] Lesson 7

[illegible]

	a	b	c	d	e
# 11	46 +10	[illegible]	26 +30	[illegible]	56 +40
# 12	45 +11	[illegible]	25 +31	[illegible]	56 +41
# 13	44 +12	[illegible]	[illegible]	[illegible]	56 +42
# 14	43 +13	[illegible]	[illegible]	[illegible]	[illegible] +43
# 15	42 +14	[illegible]	[illegible]	[illegible]	[illegible] +44
# 16	41 +15	[illegible]	[illegible]	[illegible]	56 +45
# 17	40 +16	[illegible]	[illegible]	[illegible]	56 +46
# 18	39 +17	[illegible]	[illegible]	[illegible]	56 +47
# 19	38 +18	[illegible]	[illegible]	[illegible]	56 +48
# 20	37 +19	[illegible]	[illegible]	[illegible]	56 +49

Grade 2 Chapter 2 Lesson 8
Number 57
5 tens – 1 ten = 4 tens   1 ten = 10 ones

a)	b)	c)	d)
1	4 17	1	4 17
29	5 7	19	57
+28	- 2 8	+38	- 19
57	2 9	57	38

Add or Subtract numbers in column

	a	b	c	d	e
# 1	57 +0	57 -0	37 +20	57 -20	17 +40
# 2	56 +1	57 -1	36 +21	57 -21	16 +41
# 3	55 +2	57 -2	35 +22	57 -22	15 +42
# 4	54 +3	57 -3	34 +23	57 -23	14 +43
# 5	53 +4	57 -4	33 +24	57 -24	13 +44
# 6	52 +5	57 -5	32 + 25	57 -25	12 +45
# 7	51 +6	57 -6	31 +26	57 -26	11 +46
# 8	50 +7	57 -7	30 +27	57 -27	10 +47
# 9	49 +8	57 -8	29 +28	57 -28	9 +48
# 10	48 +9	57 -9	28 +29	57 -29	8 +49

Grade 2 Chapter 2 Lesson 8
Number 57
Add or Subtract

	a	b	c	d	e
# 11	47 +10	57 -10	27 +30	57 -30	57 -40
# 12	46 +11	57 -11	26 +31	57 -31	57 -41
# 13	45 +12	57 -12	25 +32	57 -32	57 -42
# 14	44 +13	57 -13	24 +33	57 -33	57 -43
# 15	43 +14	57 -14	23 +34	57 -34	57 -44
# 16	42 +15	57 -15	22 +35	57 -35	57 -45
# 17	41 +16	57 -16	21 +36	57 -36	57 -46
# 18	40 +17	57 -17	20 +37	57 -37	57 -47
# 19	39 +18	57 -18	19 +38	57 -38	57 -48
# 20	38 +19	57 -19	18 +39	57 -39	57 -49

Grade 2 Chapter 2 Lesson 9
Number 58
5 tens – 1 ten = 4 tens   1 ten = 10 ones

a)	b)	c)	d)
1	4 18	1	4 18
29	5 8	19	58
+29	- 2 9	+39	- 19
58	2 9	58	39

Add or Subtract numbers in column

	a	b	c	d	e
# 1	58 +0	58 -0	38 +20	58 -20	18 +40
# 2	57 +1	58 -1	37 +21	58 -21	17 +41
# 3	56 +2	58 -2	36 +22	58 -22	16 +42
# 4	55 +3	58 -3	35 +23	58 -23	15 +43
# 5	54 +4	58 -4	34 +24	58 -24	14 +44
# 6	53 +5	58 -5	33 + 25	58 -25	13 +45
# 7	52 +6	58 -6	32 +26	58 -26	12 +46
# 8	51 +7	58 -7	31 +27	58 -27	11 +47
# 9	50 +8	58 -8	30 +28	58 -28	10 +48
# 10	49 +9	58 -9	29 +29	58 -29	9 +49

[illegible] son 9

[illegible]

	a		c	
# 11	48 +10	[illegible]	28 +30	[illegible]
# 12	47 +11	[illegible]	27 +31	[illegible]
# 13	46 +12	[illegible]	26 +32	[illegible]
# 14	45 +13	[illegible]	25 +33	[illegible]
# 15	44 +14	[illegible]	24 +34	[illegible]
# 16	43 +15	[illegible]	23 +35	[illegible]
# 17	42 +16	[illegible]	22 +36	[illegible]
# 18	41 +17	[illegible]	21 +37	[illegible]
# 19	40 +18	[illegible]	20 +38	[illegible]
# 20	39 +19	[illegible]	19 +39	[illegible]

Grade 2 Chapter 2 Lesson 10
Number 59

a)	b)	c)	d)
29	59	19	59
+ 30	- 30	+40	-19
59	29	59	40

Add or Subtract numbers in column

	a	b	c	d	e
# 1	59 +0	59 -0	39 +20	59 -20	19 +40
# 2	58 +1	59 -1	38 +21	59 -21	18 +41
# 3	57 +2	59 -2	37 +22	59 -22	17 +42
# 4	56 +3	59 -3	36 +23	59 -23	16 +43
# 5	55 +4	59 -4	35 +24	59 -24	15 +44
# 6	54 +5	59 -5	34 + 25	59 -25	14 +45
# 7	53 +6	59 -6	33 +26	59 -26	13 +46
# 8	52 +7	59 -7	32 +27	59 -27	12 +47
# 9	51 +8	59 -8	31 +28	59 -28	11 +48
# 10	50 +9	59 -9	30 +29	59 -29	10 +49

Grade 2 Chapter 2 Lesson 10
Number 59
Add or Subtract

	a	b	c	d	e
# 11	49 +10	59 -10	29 +30	59 -30	59 -40
# 12	48 +11	59 -11	28 +31	59 -31	59 -41
# 13	47 +12	59 -12	27 +32	59 -32	59 -42
# 14	46 +13	59 -13	26 +33	59 -33	59 -43
# 15	45 +14	59 -14	25 +34	59 -34	59 -44
# 16	44 +15	59 -15	24 +35	59 -35	59 -45
# 17	43 +16	59 -16	23 +36	59 -36	59 -46
# 18	42 +17	59 -17	22 +37	59 -37	59 -47
# 19	41 +18	59 -18	21 +38	59 -38	59 -48
# 20	40 +19	59 -19	20 +39	59 -39	59 -49

Grade 2 Chapter 2 Lesson 11
Numbers from 50 to 54
Test 2 A
Add or Subtract

	a	b	c	d	e
# 1	49 + 1	50 - 1	29 +21	50 - 21	9 + 41
# 2	39 +11	50 - 11	19 + 31	50 - 31	50 - 41
# 3	48 + 3	51 - 3	28 +23	51 - 23	8 +43
# 4	38 + 13	51 -13	18 +33	51 -33	51 -43
# 5	47 +5	52 -5	27 +25	52 -25	7 +45
# 6	37 +15	52 -15	17 +35	52 -35	52 -45
# 7	46 + 7	53 - 7	26 +27	53 -27	6 +47
# 8	36 + 17	53 - 17	16 +37	53 -37	53 -47
# 9	45 +9	54 -9	25 +29	54 -29	7 +47
# 10	35 +19	54 -19	15 +39	54 -39	54 -49

Gr[illegible] Chap[illegible]son 11
[illegible]rs [illegible] 59
Tes[illegible]

	a	b	c	d	e
# 1	45 +10	[illegible]	25 [illegible]30		55 -40
# 2	44 + 11	[illegible]	24 [illegible]31		55 -41
# 3	44 +12	[illegible]	24 [illegible]2		56 -42
# 4	43 +13	[illegible]	[illegible]3 [illegible]3		56 -43
# 5	43 +14	[illegible]	[illegible]		57 -44
# 6	42 +15	[illegible]	[illegible]		57 -45
# 7	42 +16	[illegible]	[illegible]		58 -46
# 8	41 +17	[illegible]	[illegible]		58 -47
# 9	41 +18	[illegible]	[illegible]	[illegible]	59 -48
# 10	40 +19	[illegible]	[illegible]	[illegible]	59 -49

Grade 2 Answers Chapter 2

Lesson 1

	a, b, c, d, e		a, b, c, d, e
# 1	50, 50, 50, 30, 50	# 11	50, 40, 50, 20, 10
# 2	50, 49, 50, 29, 50	# 12	50, 39, 50, 19, 9
# 3	50, 48, 50, 28, 50	# 13	50, 38, 50, 18, 8
# 4	50, 47, 50, 27, 50	# 14	50, 37, 50, 17, 7
# 5	50, 46, 50, 26, 50	# 15	50, 36, 50, 16, 6
# 6	50, 45, 50, 25, 50	# 16	50, 35, 50, 15, 5
# 7	50, 44, 50, 24, 50	# 17	50, 34, 50, 14, 4
# 8	50, 43, 50, 23, 50	# 18	50, 33, 50, 13, 3
# 9	50, 42, 50, 22, 50	# 19	50, 32, 50, 12, 2
# 10	50, 41, 50, 21, 50	# 20	50, 31, 50, 11, 1

Lesson 2

	a, b, c, d, e		a, b, c, d, e
# 1	51, 51, 51, 31, 51	# 11	51, 41, 51, 21, 11
# 2	51, 50, 51, 30, 51	# 12	51, 40, 51, 20, 10
# 3	51, 49, 51, 29, 51	# 13	51, 39, 51, 19, 9
# 4	51, 48, 51, 28, 51	# 14	51, 38, 51, 18, 8
# 5	51, 47, 51, 27, 51	# 15	51, 37, 51, 17, 7
# 6	51, 46, 51, 26, 51	# 16	51, 36, 51, 16, 6
# 7	51, 45, 51, 25, 51	# 17	51, 35, 51, 15, 5
# 8	51, 44, 51, 24, 51	# 18	51, 34, 51, 14, 4
# 9	51, 43, 51, 23, 51	# 19	51, 33, 51, 13, 3
# 10	51, 42, 51, 22, 51	# 20	51, 32, 51, 12, 2

Lesson 3

	a, b, c, d, e		a, b, c, d, e
# 1	52, 52, 52, 32, 52	# 11	52, 42, 52, 22, 12
# 2	52, 51, 52, 31, 52	# 12	52, 41, 52, 21, 11
# 3	52, 50, 52, 30, 52	# 13	52, 40, 52, 20, 10
# 4	52, 49, 52, 29, 52	# 14	52, 39, 52, 19, 9
# 5	52, 48, 52, 28, 52	# 15	52, 38, 52, 18, 8
# 6	52, 47, 52, 27, 52	# 16	52, 37, 52, 17, 7
# 7	52, 46, 52, 26, 52	# 17	52, 36, 52, 16, 6
# 8	52, 45, 52, 25, 52	# 18	52, 35, 52, 15, 5
# 9	52, 44, 52, 24, 52	# 19	52, 34, 52, 14, 4
# 10	52, 43, 52, 23, 52	# 20	52, 33, 52, 13, 3

Grade 2 Answers Chapter 2

Lesson 4

	a b c d e		a b c d e
# 1	53, 53, 53, 33, 53	# 11	53, 43, 53, 23, 13
# 2	53, 52, 53, 32, 53	# 12	53, 42, 53, 22, 12
# 3	53, 51, 53, 31, 53	# 13	53, 41, 53, 21, 11
# 4	53, 50, 53, 30, 53	# 14	53, 40, 53, 20, 10
# 5	53, 49, 53, 29, 53	# 15	53, 39, 53, 19, 9
# 6	53, 48, 53, 28, 53	# 16	53, 38, 53, 18, 8
# 7	53, 47, 53, 27, 53	# 17	53, 37, 53, 17, 7
# 8	53, 46, 53, 26, 53	# 18	53, 36, 53, 16, 6
# 9	53, 45, 53, 25, 53	# 19	53, 35, 53, 15, 5
# 10	53, 44, 53, 24, 53	# 20	53, 34, 53, 14, 4

Lesson 5

	a b c d e		a b c d e
# 1	54, 54, 54, 34, 54	# 11	54, 44, 54, 24, 14
# 2	54, 53, 54, 33, 54	# 12	54, 43, 54, 23, 13
# 3	54, 52, 54, 32, 54	# 13	54, 42, 54, 22, 12
# 4	54, 51, 54, 31, 54	# 14	54, 41, 54, 21, 11
# 5	54, 50, 54, 30, 54	# 15	54, 40, 54, 20, 10
# 6	54, 49, 54, 29, 54	# 16	54, 39, 54, 19, 9
# 7	54, 48, 54, 28, 54	# 17	54, 38, 54, 18, 8
# 8	54, 47, 54, 27, 54	# 18	54, 37, 54, 17, 7
# 9	54, 46, 54, 26, 54	# 19	54, 36, 54, 16, 6
# 10	54, 45, 54, 25, 54	# 20	54, 35, 54, 15, 5

Lesson 6

	a b c d e		a b c d e
# 1	55, 55, 55, 35, 55	# 11	55, 45, 55, 25, 15
# 2	55, 54, 55, 34, 55	# 12	55, 44, 55, 24, 14
# 3	55, 53, 55, 33, 55	# 13	55, 43, 55, 23, 13
# 4	55, 52, 55, 32, 55	# 14	55, 42, 55, 22, 12
# 5	55, 51, 55, 31, 55	# 15	55, 41, 55, 21, 11
# 6	55, 50, 55, 30, 55	# 16	55, 40, 55, 20, 10
# 7	55, 49, 55, 29, 55	# 17	55, 39, 55, 19, 9
# 8	55, 48, 55, 28, 55	# 18	55, 38, 55, 18, 8
# 9	55, 47, 55, 27, 55	# 19	55, 37, 55, 17, 7
# 10	55, 46, 55, 26, 55	# 20	55, 36, 55, 16, 6

Grade 2 Answers Chapter 2

Lesson 7

	a, b, c, d, e
# 1	56, 56, 56, 36, 56
# 2	56, 55, 56, 35, 56
# 3	56, 54, 56, 34, 56
# 4	56, 53, 56, 33, 56
# 5	56, 52, 56, 32, 56
# 6	56, 51, 56, 31, 56
# 7	56, 50, 56, 30, 56
# 8	56, 49, 56, 29, 56
# 9	56, 48, 56, 28, 56
# 10	56, 47, 56, 27, 56
# 11	56, 46, 56, 26, 16
# 12	56, 45, 56, 25, 15
# 13	56, 44, 56, 24, 14
# 14	56, 43, 56, 23, 13
# 15	56, 42, 56, 22, 12
# 16	56, 41, 56, 21, 11
# 17	56, 40, 56, 20, 10
# 18	56, 39, 56, 19, 9
# 19	56, 38, 56, 18, 8
# 20	56, 37, 56, 17, 7

Lesson 8

	a, b, c, d, e
# 1	57, 57, 57, 37, 57
# 2	57, 56, 57, 36, 57
# 3	57, 55, 57, 35, 57
# 4	57, 54, 57, 34, 57
# 5	57, 53, 57, 33, 57
# 6	57, 52, 57, 32, 57
# 7	57, 51, 57, 31, 57
# 8	57, 50, 57, 30, 57
# 9	57, 49, 57, 29, 57
# 10	57, 48, 57, 28, 57
# 11	57, 47, 57, 27, 17
# 12	57, 46, 57, 26, 16
# 13	57, 45, 57, 25, 15
# 14	57, 44, 57, 24, 14
# 15	57, 43, 57, 23, 13
# 16	57, 42, 57, 22, 12
# 17	57, 41, 57, 21, 11
# 18	57, 40, 57, 20, 10
# 19	57, 39, 57, 19, 9
# 20	57, 38, 57, 18, 8

Lesson 9

	a, b, c, d, e
# 1	58, 58, 58, 38, 58
# 2	58, 57, 58, 37, 58
# 3	58, 56, 58, 36, 58
# 4	58, 55, 58, 35, 58
# 5	58, 54, 58, 34, 58
# 6	58, 53, 58, 33, 58
# 7	58, 52, 58, 32, 58
# 8	58, 51, 58, 31, 58
# 9	58, 50, 58, 30, 58
# 10	58, 49, 58, 29, 58
# 11	58, 48, 58, 28, 18
# 12	58, 47, 58, 27, 17
# 13	58, 46, 58, 26, 16
# 14	58, 45, 58, 25, 15
# 15	58, 44, 58, 24, 14
# 16	58, 43, 58, 23, 13
# 17	58, 42, 58, 22, 12
# 18	58, 41, 58, 21, 11
# 19	58, 40, 58, 20, 10
# 20	58, 39, 58, 19, 9

Lesson 10

a b c

\# 1 59, 59, 59, [illegible] 49, [illegible]
\# 2 59, 58, 59, [illegible] 48, [illegible]
\# 3 59, 57, 59, [illegible] 47, [illegible]
\# 4 59, 56, 59, [illegible] 46, [illegible]
\# 5 59, 55, 59, [illegible] 45, [illegible]
\# 6 59, 54, 59, [illegible] 44, [illegible]
\# 7 59, 53, 59, [illegible] 43, [illegible]
\# 8 59, 52, 59, [illegible] 42, [illegible]
\# 9 59, 51, 59, [illegible] 41, [illegible]
\# 10 59, 50, 59, [illegible] 40, [illegible]

Lesson 11 Test [illegible]

Test 2 A [illegible]

	a	b	[illegible]	c	d	e
# 1	50	49	[illegible]	55	25	15
# 2	50	39	[illegible]	55	24	14
# 3	51	48	[illegible]	56	24	14
# 4	51	38	[illegible]	56	23	13
# 5	52	47	[illegible]	57	23	13
# 6	52	37	[illegible]	57	22	12
# 7	53	46	[illegible]	58	22	12
# 8	53	36	[illegible]	58	21	11
# 9	54	45	[illegible]	59	21	11
# 10	54	35	[illegible]	59	20	10

Grade 2

Chapter 3 Addition or Subtraction Numbers (30-39)

Grade 2 Chapter 3 Contents
Addition or Subtraction Numbers (30-39)
Lesson 1 - 11
Basic Problems

Work with lesson 1, read rules and examples, 2) Solve the problems from contents, 3) Check answers in the end of chapter 1, 4) Work with the next lessons 2, 3, 4, 5, …, 5) Solve Test 1 A, 6) Grading Scale: A =10 points, B=9points, C=8 points, D=7 points, F=0-6 points, 7) If you solve 10 problems correctly, you get an A, 8) If have an A, you can move on to next chapter, 9) If your grade is lower than an A, work with the mistakes, then solve Test 1B, 10) When you finish all of the chapters, solve Final Exam in last chapter.

Grade 2 Chapter 3 Lesson 1
Number 30
Add or Subtract numbers in column

3 tens – 1 ten = 2 tens   1 ten = 10 ones

a)	b)	c)	d)
1	2 10	1	2 10
19	3 0	18	30
+11	- 1 1	+12	- 12
30	1 9	30	18

Add or Subtract

	a	b	c	d	e
# 1	30 + 0	30 - 0	20 + 10	30 - 10	10 + 20
# 2	29 + 1	30 - 1	19 +11	30 - 11	9 + 21
# 3	28 + 2	30 - 2	18 + 12	30 - 12	8 + 22
# 4	27 + 3	30 - 3	17 + 13	30 - 13	7 + 23
# 5	26 + 4	30 - 4	16 + 14	30 - 14	6 + 24
# 6	25 + 5	30 - 5	15 + 15	30 - 15	5 + 25
# 7	24 + 6	30 - 6	14 + 16	30 - 16	4 + 26
# 8	23 + 7	30 - 7	13 + 17	30 - 17	3 + 27
# 9	22 + 8	30 - 8	12 + 18	30 - 18	2 + 28
# 10	21 + 9	30 - 9	11 + 19	30 - 19	1 + 29

Gr[illegible] Chap[illegible] [illegible]sson 2

[illegible]

Add [illegible] in column

[illegible] ns – [illegible] ten [illegible] = 10 ones

a) $\begin{array}{r} {}^{1}\\ 19 \\ +12 \\ \hline 31 \end{array}$ b) [illegible] c) [illegible]

	a	b	c	d	e
# 1	31 +0	[illegible]	[illegible]	[illegible]	[illegible]
# 2	30 +1	[illegible]	[illegible]	[illegible]	[illegible]
# 3	29 + 2	[illegible]	[illegible]	[illegible]	9 + 2
# 4	28 + 3	[illegible]	[illegible]	[illegible]	8 + 3
# 5	27 + 4	[illegible]	[illegible]	[illegible]	7 + 4
# 6	26 + 5	[illegible]	[illegible]	[illegible]	6 + 5
# 7	25 + 6	[illegible]	[illegible]	[illegible]	5 + 6
# 8	24 + 7	[illegible]	[illegible]	[illegible]	4 + 7
# 9	23 + 8	[illegible]	[illegible]	[illegible]	3 + 8
# 10	22 + 9	[illegible]	[illegible]	[illegible]	2 + 9

Grade 2 Chapter 3 Lesson 3

Number 32

3 tens – 1 ten = 2 tens   1 ten = 10 ones

a)	b)	c)	d)
1	2 12	1	2 12
19	3 2	18	32
+13	- 1 3	+14	- 18
32	1 9	32	14

Add or Subtract numbers in column

	a	b	c	d	e
# 1	32	32	22	32	12
	+0	-0	+10	-10	+20
# 2	31	32	21	32	11
	+1	-1	+11	-11	+21
# 3	30	32	20	32	10
	+2	-2	+12	- 12	+22
# 4	29	32	19	32	9
	+3	-3	+13	-13	+23
# 5	28	32	18	32	8
	+4	-4	+14	-14	+24
# 6	27	32	17	32	7
	+5	-5	+15	-15	+25
# 7	26	32	16	32	6
	+6	-6	+16	-16	+26
# 8	25	32	15	32	5
	+7	- 7	+17	- 17	+27
# 9	24	32	14	32	4
	+8	- 8	+18	-18	+28
# 10	23	32	13	32	3
	+9	- 9	+19	-19	+29

Grade 2 Chapter 3 Lesson 4
Number 33

3 tens – 1 ten = 2 tens   1 ten = 10 ones

a)	b)	c)	d)
1	2 13	1	2 13
19	3 3	18	33
+14	- 1 4	+15	- 18
33	2 9	33	15

Add or Subtract numbers in column

	a	b	c	d	e
# 1	33 + 0	33 - 0	23 +10	33 -10	13 +20
# 2	32 + 1	33 -1	22 +11	33 -11	12 +21
# 3	31 + 2	33 - 2	21 +12	33 -12	11 +22
# 4	30 + 3	33 - 3	20 +13	33 -13	10 +23
# 5	29 + 4	33 - 4	19 +14	33 -14	9 +24
# 6	28 + 5	33 - 5	18 +15	33 -15	8 +25
# 7	27 + 6	33 - 6	17 +26	33 -16	7 +26
# 8	26 + 7	33 - 7	16 +17	33 -17	6 +27
# 9	25 + 8	33 - 8	15 +18	33 -18	5 +28
# 10	24 +9	33 - 9	14 +19	33 -19	4 +29

Grade 2 Chapter 3 Lesson 5
Number 34
Add or Subtract numbers in column
3 tens – 1 ten = 2 tens   1 ten = 10 ones

a)	b)	c)	d)
1	2 14	1	2 14
19	3 4	18	34
+15	- 1 5	+16	- 19
34	1 9	34	15

Add or Subtract numbers in column

	a	b	c	d	e
# 1	34 +0	34 -0	24 +10	34 -10	14 +20
# 2	33 +1	34 -1	23 +11	34 -11	13 +21
# 3	32 +2	34 -2	22 +12	34 -12	12 +22
# 4	31 +3	34 -3	21 +13	34 -13	11 +23
# 5	30 +4	34 -4	20 +14	34 -14	10 +24
# 6	29 +5	34 -5	19 +15	34 -15	9 +25
# 7	28 +6	34 -6	18 +16	34 -16	8 +26
# 8	27 +7	34 -7	17 +17	34 -17	7 +27
# 9	26 +8	34 -8	16 +18	34 -18	6 +28
# 10	25 +9	34 -9	15 +19	34 -19	7 +27

[illegible] Chap[illegible] [illegible]son 6

[illegible]

[illegible] tens [illegible] 2 [illegible] = 10 ones

a)
$$\begin{array}{r} ^{1}\ \ \\ 19 \\ +16 \\ \hline 35 \end{array}$$

b) [illegible]

c) [illegible]

[illegible] Add [illegible] in column

	a	b	c	d	e
# 1	35 +0	[illegible]	[illegible]	[illegible]	15 20
# 2	34 +1	[illegible]	[illegible]	[illegible]	14 21
# 3	33 +2	[illegible]	[illegible]	[illegible]	13 22
# 4	32 +3	[illegible]	[illegible]	[illegible]	12 23
# 5	31 +4	[illegible]	[illegible]	[illegible]	11 24
# 6	30 +5	[illegible]	[illegible]	[illegible]	10 25
# 7	29 +6	[illegible]	[illegible]	[illegible]	9 26
# 8	28 +7	[illegible]	[illegible]	[illegible]	8 27
# 9	27 +8	[illegible]	[illegible]	[illegible]	7 28
# 10	26 +9	[illegible]	[illegible]	[illegible]	6 29

Grade 2 Chapter 3 Lesson 7
Number 36
Add or Subtract numbers in column
3 tens – 1 ten = 2 tens   1 ten = 10 ones

a)	b)	c)	d)
1	2 16	1	2 16
19	3 6	18	36
+17	- 1 7	+18	- 18
36	1 9	36	18

Add or Subtract numbers in column

	a	b	c	d	e
# 1	36	36	26	36	16
	+0	-0	+10	-10	+20
# 2	35	36	25	36	15
	+1	-1	+11	-11	+21
# 3	34	36	24	36	14
	+2	-2	+12	-12	+22
# 4	33	36	23	36	13
	+3	-3	+13	-13	+23
# 5	32	36	22	36	12
	+4	-4	+14	-14	+24
# 6	31	36	21	36	11
	+5	-5	+ 15	-15	+25
# 7	30	36	20	36	10
	+6	-6	+16	-16	+26
# 8	29	36	19	36	9
	+7	-7	+17	-17	+27
# 9	28	36	18	36	8
	+8	-8	+18	-18	+28
# 10	27	36	17	36	7
	+9	-9	+19	-19	+ 29

Grade 2 Chapter 3 Lesson 8
Number 37
3 tens – 1 ten = 2 tens   1 ten = 10 ones

a)	b)	c)	d)
1	2 17	1	2 17
19	3 7	18	37
+18	- 19	+ 19	- 18
37	18	37	19

Add or Subtract numbers in column

	a	b	c	d	e
# 1	37 +0	37 -0	27 +10	37 -10	17 +20
# 2	36 +1	37 -1	26 +11	37 -11	16 +21
# 3	35 +2	37 -2	25 +12	37 -12	15 +22
# 4	34 +3	37 -3	24 +13	37 -13	14 +23
# 5	33 +4	37 -4	23 +14	37 -14	13 +24
# 6	32 +5	37 -5	22 + 15	37 -15	12 +25
# 7	31 +6	37 -6	21 +16	37 -16	11 +26
# 8	30 +7	37 -7	20 +17	37 -17	10 +27
# 9	29 +8	37 -8	19 +18	37 -18	9 +28
# 10	28 +9	37 -9	18 +19	37 -19	8 +29

Grade 2 Chapter 3 Lesson 9
Number 38

3 tens – 1 ten = 2 tens   1 ten = 10 ones

a)	b)	c)	d)
1	2 18	1	
19	3 8	18	38
+19	- 1 9	+20	- 18
38	1 9	38	20

Add or Subtract numbers in column

	a	b	c	d	e
# 1	38 +0	38 -0	28 +10	38 -10	18 +20
# 2	37 +1	38 -1	27 +11	38 -11	17 +21
# 3	36 +2	38 -2	26 +12	38 -12	16 +22
# 4	35 +3	38 -3	25 +13	38 -13	15 +23
# 5	34 +4	38 -4	24 +14	38 -14	14 +24
# 6	33 +5	38 -5	23 + 15	38 -15	13 +25
# 7	32 +6	38 -6	22 +16	38 -16	12 +26
# 8	31 +7	38 -7	21 +17	38 -17	11 +27
# 9	30 +8	38 -8	20 +18	38 -18	10 +28
# 10	29 +9	38 -9	19 +19	38 -19	9 +29

[illegible] Gr[illegible] [illegible] Chap[illegible] [illegible]son 10

[illegible] Column

a) [illegible] b) 39 [illegible] d) 39

a)	b)	c)	d)
[illegible]	39	[illegible]	39
[illegible]	- 10	[illegible]	-18
[illegible]	19	[illegible]	21

Add [illegible] [illegible]ct [illegible] in column

	a	b	c	d
# 1	39 +0	[illegible]	29 +10	[illegible]
# 2	38 +1	[illegible]	28 +11	[illegible]
# 3	37 +2	[illegible]	27 +12	[illegible]
# 4	36 +3	[illegible]	26 +13	[illegible]
# 5	35 +4	[illegible]	25 +14	[illegible]
# 6	34 +5	[illegible]	24 + 15	[illegible]
# 7	33 +6	[illegible]	23 +16	[illegible]
# 8	32 +7	[illegible]	22 +17	[illegible]
# 9	31 +8	[illegible]	21 +18	[illegible]
# 10	30 +9	[illegible]	20 +19	[illegible]

Grade 2 Chapter 3 Lesson 11
Numbers from 30 to 39
Test 3 A
Add or Subtract

	a	b
# 1	30 + 1	30 - 1
# 2	30 + 2	30 - 2
# 3	31 + 3	31 - 3
# 4	31 + 4	31 - 4
# 5	32 +5	32 -5
# 6	32 + 1	32 - 1
# 7	33 + 2	33 - 2
# 8	33 + 3	33 - 3
# 9	34 + 4	34 - 4
# 10	34 + 5	34 - 5

Grade 2 Chapter 3 Lesson 11
Numbers from 30 to 39
Test 3 B

	a	b
# 1	30 + 9	30 - 9
# 2	30 + 8	30 - 8
# 3	31 + 7	31 - 7
# 4	31 + 6	31 - 6
# 5	32 + 5	32 - 5
# 6	32 + 6	32 - 6
# 7	33 + 6	33 - 6
# 8	33 + 5	33 - 5
# 9	34 + 5	34 - 5
# 10	35 + 4	35 - 4

Grade 2 Answers Chapter 3

Lesson 1 Number 30

	a, b, c, d, e
# 1	30, 30, 30, 20, 30
# 2	30, 29, 30, 19, 30
# 3	30, 28, 30, 18, 30
# 4	30, 27, 30, 17, 30
# 5	30, 26, 30, 16, 30
# 6	30, 25, 30, 15, 30
# 7	30, 24, 30, 14, 30
# 8	30, 23, 30, 13, 30
# 9	30, 22, 30, 12, 30
# 10	30, 21, 30, 11, 30

Lesson 2 Number 31

	a, b, c, d, e
# 1	31, 31, 31, 21, 31
# 2	31, 30, 31, 20, 31
# 3	31, 29, 31, 19, 31
# 4	31, 28, 31, 18, 31
# 5	31, 27, 31, 17, 31
# 6	31, 26, 31, 16, 31
# 7	31, 25, 31, 15, 31
# 8	31, 24, 31, 14, 31
# 9	31, 23, 31, 13, 31
# 10	31, 22, 31, 12, 31

Lesson 3 Number 32

	a, b, c, d, e
# 1	32, 32, 32, 22, 32
# 2	32, 31, 32, 21, 32
# 3	32, 30, 32, 20, 32
# 4	32, 29, 32, 19, 32
# 5	32, 28, 32, 18, 32
# 6	32, 27, 32, 17, 32
# 7	32, 26, 32, 16, 32
# 8	32, 25, 32, 15, 32
# 9	32, 24, 32, 14, 32
# 10	32, 23, 32, 13, 32

Lesson 4 Numb

a b c

# 1	33, 33, 33
# 2	33, 32, 33
# 3	33, 31, 33
# 4	33, 30, 33
# 5	33, 29, 33
# 6	33, 28, 33
# 7	33, 27, 33
# 8	33, 26, 33
# 9	33, 25, 33
# 10	33, 24, 33

Lesson 5 Numb

a b c

# 1	34, 34, 34
# 2	34, 33, 34
# 3	34, 32, 34
# 4	34, 31, 34
# 5	34, 30, 34
# 6	34, 29, 34
# 7	34, 28, 34
# 8	34, 27, 34
# 9	34, 26, 34
# 10	34, 25, 34

Lesson 6 Numb

a b c

# 1	35, 35, 35
# 2	35, 34, 35
# 3	35, 33, 35
# 4	35, 32, 35
# 5	35, 31, 35
# 6	35, 30, 35
# 7	35, 29, 35
# 8	35, 28, 35
# 9	35, 27, 35
# 10	35, 26, 35

Grade 2 Answers Chapter 3

Lesson 7 Number 36

	a	b	c	d	e
# 1	36,	36,	36,	26,	36
# 2	36,	35,	36,	25,	36
# 3	36,	34,	36,	24,	36
# 4	36,	33,	36,	23,	36
# 5	36,	32,	36,	22,	36
# 6	36,	31,	36,	21,	36
# 7	36,	30,	36,	20,	36
# 8	36,	29,	36,	19,	36
# 9	36,	28,	36,	18,	36
# 10	36,	27,	36,	17,	36

Lesson 8 Number 37

	a	b	c	d	e
# 1	37,	37,	37,	27,	37
# 2	37,	36,	37,	26,	37
# 3	37,	35,	37,	25,	37
# 4	37,	34,	37,	24,	37
# 5	37,	33,	37,	23,	37
# 6	37,	32,	37,	22,	37
# 7	37,	31,	37,	21,	37
# 8	37,	30,	37,	20,	37
# 9	37,	29,	37,	19,	37
# 10	37,	28,	37,	18,	37

Lesson 9 Number 38

	a	b	c	d	e
# 1	38,	38,	38,	28,	38
# 2	38,	37,	38,	27,	38
# 3	38,	36,	38,	26,	38
# 4	38,	35,	38,	25,	38
# 5	38,	34,	38,	24,	38
# 6	38,	33,	38,	23,	38
# 7	38,	32,	38,	22,	38
# 8	38,	31,	38,	21,	38
# 9	38,	30,	38,	20,	38
# 10	38,	29,	38,	19,	38

Lesson 10 Number 39

	a	b	c	d	e
# 1	39,	39,	39,	19,	39
# 2	39,	38,	39,	18,	39
# 3	39,	37,	39,	17,	39
# 4	39,	36,	39,	16,	39
# 5	39,	35,	39,	15,	39
# 6	39,	34,	39,	14,	39
# 7	39,	33,	39,	13,	39
# 8	39,	32,	39,	12,	39
# 9	39,	31,	39,	11,	39
# 10	39,	30,	39,	10,	39

Lesson 11 Test 3

Test 3 A

	a	b
# 1	31	29
# 2	32	28
# 3	34	28
# 4	35	27
# 5	37	27
# 6	33	31
# 7	35	31
# 8	36	30
# 9	38	30
# 10	39	31

Test 3 B

	a	b
# 1	39	21
# 2	38	22
# 3	38	24
# 4	37	25
# 5	37	27
# 6	38	26
# 7	39	27
# 8	38	28
# 9	39	29
# 10	39	31

Grade 2

Chapter 4

Finding unknowns (40–49)

Gr[illegible] Chap[illegible]ntents
Fi[illegible]-49)
[illegible]esson[illegible]
[illegible]nced[illegible]s

Work with lesso[illegible] rules[illegible]ve the problems from contents, 3) Ch[illegible]rs in [illegible] 4) Work with the next lessons 2, 3, 4, 5[illegible]lve T[illegible]cale: A =10 points, B=9points, C=8[illegible]=7 p[illegible]-6 [illegible] If you solve 10 problems correctly, you g[illegible]) If h[illegible] yo[illegible]ve on to next chapter, 9) If your grad[illegible]han [illegible]rk [illegible]istakes, then solve Test 1B, 10) When you fi[illegible] the [illegible], sol[illegible]Exam in last chapter.

Lesson 1 Summ[illegible]ve th[illegible]s # [illegible], 20
Lesson 2 Summ[illegible]view [illegible] 1 B # 1, 2, 3, 4
Lesson 3 Summ[illegible]ve th[illegible]s #[illegible]
Lesson 4 Summ[illegible]d or [illegible] Num[illegible]Row
Lesson 5 Numb[illegible]e th[illegible]s # [illegible] d, e), # 16 (a, b, c, d, e)
Lesson 6 Numb[illegible]e th[illegible]s # [illegible] d, e), # 16 (a, b, c, d, e)
Lesson 7 Numb[illegible]e th[illegible]s # [illegible] d, e), # 16 (a, b, c, d, e)
Lesson 8 Numb[illegible]e th[illegible]s # [illegible] d, e), # 16 (a, b, c, d, e)

Lesson 9 Numb[illegible]e th[illegible]s # [illegible] d, e), # 16 (a, b, c, d, e)
Lesson 10 Num[illegible]lve th[illegible]ms [illegible]c, d, e), # 16 (a, b, c, d, e)
Lesson 11 Num[illegible]lve th[illegible]ms [illegible]c, d, e), # 16 (a, b, c, d, e)
Lesson 12 Num[illegible]lve th[illegible]ms [illegible]c, d, e), # 16 (a, b, c, d, e)
Lesson 13 Num[illegible]lve th[illegible]ms [illegible]c, d, e), # 16 (a, b, c, d, e)
Lesson 14 Num[illegible]lve th[illegible]ms [illegible]c, d, e), # 16 (a, b, c, d, e)
Lesson 15 Less[illegible] 4 for [illegible] tes[illegible]lems)
Solve Test 4 A [illegible] d, e[illegible](a, [illegible]
Solve Test 4 B [illegible] d, e[illegible](a, [illegible]
<u>Lesson 16 (Basi[illegible])</u>
Solve Review 2 [illegible], 7, 9[illegible] 5, 1[illegible]
Solve Review 2 [illegible], 7, 9[illegible] 5, 1[illegible]
<u>Lesson 17 (Basi[illegible])</u>
Solve Final Exa[illegible], 3, 5, [illegible] 13, [illegible]
Solve Final Exa[illegible] 3, 5, [illegible] 3, 1[illegible]
Answers Chapt[illegible]

Grade 2 Chapter 4 Lesson 1
Summary 1

Rules	Examples	
Addend + Addend = Sum	2 + 3 = 5	5 > 2 5 > 3
First Addend = Sum – Second Addend	? + 3 = 5	5 – 3 = 2
Second Addend = Sum – First Addend	2 + ? = 5	5 – 2 = 3
Minuend – Subtrahend = Difference	5 – 2 = 3	5 > 2 5 > 3
Minuend = Subtrahend + Difference	? - 2 = 3	3 + 2 = 5
Subtrahend = Minuend – Difference	5 - ? = 3	5 – 3 = 2

Put right number instead question mark

	a	b	c	d
# 1	2 + ? = 30	? + 29 = 30	0 + ? = 30	17 + ? = 30
# 2	30 – 28 = ?	30 - ? = 1	30 – ? = 0	30 – 14 = ?
# 3	3 + ? = 31	? + 29 = 31	1 + ? = 31	? + 31 = 31
# 4	31 – ? = 3	31 – 29 = ?	31 – ? = 1	31 – 31 = ?
# 5	? + 28 = 32	3 + ? = 32	? + 30 = 32	1 + ? = 32
# 6	32 – ? = 4	32 – 29 = ?	32 – ? = 2	32 – 31 = ?
# 7	? + 28 = 33	4 + ? = 33	? + 30 = 33	2 + ? = 33
# 8	33 – ? = 5	33 - 29 = ?	33 – ? = 3	33 – 31 = ?
# 9	? + 28 = 34	5 + ? = 34	? + 30 = 34	3 + ? = 34
# 10	34 – ? = 6	34 – 29 = ?	34 – ? = 4	34 - 31 = ?

Grade 2 Chapter 4 Lesson 1

Summary 1

Rules	Examples	
Addend + Addend = Sum	2 + 3 = 5	5 > 2 5 > 3
First Addend = Sum – Second Addend	? + 3 = 5	5 – 3 = 2
Second Addend = Sum – First Addend	2 + ? = 5	5 – 2 = 3
Minuend – Subtrahend = Difference	5 – 2 = 3	5 > 2 5 > 3
Minuend = Subtrahend + Difference	? - 2 = 3	3 + 2 = 5
Subtrahend = Minuend – Difference	5 - ? = 3	5 – 3 = 2

Put right number instead question mark

	a	b	c	d
# 11	7 + 28 = ?	6 + ? = 35	? + 30 = 35	4 + ? = 35
# 12	? – 28 = 7	35 – 29 = ?	35 – ? = 5	35 – 31 = ?
# 13	8 + 28 = ?	7 + ? = 36	? + 30 = 36	5 + ? = 36
# 14	? – 28 = 8	36 – ? = 7	36 – 30 = ?	36 – ? = 5
# 15	9 + 28 = ?	8 + ? = 37	? + 30 = 37	6 + ? = 37
# 16	? – 28 = 9	37 – ? = 8	37 – 30 = ?	37 – ? = 6
# 17	10 + 28 = ?	9 + ? = 38	? + 30 = 38	7 + ? = 38
# 18	? – 28 = 10	38 – ? = 9	38 – 30 = ?	38 – ? = 7
# 19	11 + 28 = ?	10 + ? = 39	? + 30 = 39	8 + ? = 39
# 20	? – 28 = 11	39 – ? = 10	39 – 30 = ?	39 – ? = 8

Grade 2 Chapter 4 Lesson 2
Summary 2
Addition

First number in addition called the addend	**2**
Second number in addition called the addend	**3**
Result of addition called the sum	**5**
Addend + Addend = Sum	**2 + 3 = 5**

First Addend = Sum – Second Addend

? + 3 = 5	**? = 5 – 3**	**? = 2**	**5 > 2**

Second Addend = Sum – First Addend

2 + ? = 5	**? = 5 – 2**	**? = 3**	**5 > 3**

Subtraction

First number in subtraction called minuend	**5**
Second number in subtraction called subtrahend	**2**
Result of subtraction called difference	**3**
Minuend – Subtrahend = Difference	**5 – 2 = 3**

Minuend = Subtrahend + Difference

? – 2 = 3	**? = 3 + 2**	**? = 5**	**5 > 3**

Subtrahend = Minuend – Difference

5 - ? = 3	**? = 5 – 3**	**? = 2**	**5 > 2**

Review 1 A
Put right number instead question mark

	a	b	c	d	e
# 1	**? + 7 = 10**	**5 + ? = 11**	**? + 5 = 12**	**5 + ? = 13**	**? + 8 = 14**
# 2	**? - 8 = 2**	**11 - ? = 2**	**? – 8 = 4**	**13 - ? = 9**	**? – 8 = 6**

Complete the tables

 # 3

	a	b	c	d	e
Addend	**?**	**9**	**9**	**?**	**11**
Addend	**6**	**?**	**8**	**8**	**?**
Sum	**15**	**16**	**?**	**18**	**19**

e) 11 + ? = 19

 # 4

	a	b	c	d	e
Minuend	**?**	**16**	**17**	**?**	**19**
Subtrahend	**8**	**?**	**8**	**8**	**?**
Difference	**7**	**8**	**?**	**10**	**9**

e) 19 - ? = 9

[illegible] son 2

[illegible]

[illegible] rst [illegible] nd Addend

? [illegible] ? = 2

[illegible] econ [illegible] st Addend

2 [illegible] ? = 3

[illegible] inu [illegible] btr [illegible] ifference

[illegible] ? = 5

[illegible] ubt [illegible] Min [illegible] ifference

5 [illegible] ? = 2

[illegible]

[illegible] rig [illegible] estion mark

			d	e
# 1	? [illegible]	4 + [illegible]	4 + ? = 13	? + 9 = 14
# 2	? [illegible]	11 [illegible]	13 - ? = 8	? – 9 = 5

[illegible] plete [illegible] s

[illegible] 3	b	[illegible]	e
[illegible] nd	8	?	9
[illegible] nd	?	9	?
[illegible]	16	18	19

e) 9 + ? = 19

[illegible] 4	b	d	e
[illegible] end	16	?	19
[illegible] hend	?	9	?
[illegible] nce	9	9	8

e) 19 - ? = 8

Grade 2 Chapter 4 Lesson 3
Summary 3 Examples

a) ? + 3 = 5   5 – 3 = 2 c) 2 + ? = 5   5 – 2 = 3
b) ? - 2 = 3   3 + 2 = 5 d) 5 - ? = 3   5 – 3 = 2

Complete the tables

Number 40

 # 1 a) ? + 7 = 40

# 1	a	b	c	d	e
Addend	?	25	17		1
Addend	7		23	31	
Sum	40	40		40	40

 # 2 a) ? - 7 = 33

# 2	a				
Minuend	?	40	40		40
Subtrahend	7		23	31	
Difference	33	25		9	1

Number 41

 # 3 b) 26 + ? = 41

# 3		b			
Addend		26	18		2
Addend	7	?	23	31	
Sum	41	41		41	41

 # 4 b) 41 - ? = 26

# 4		b			
Minuend		41	41		41
Subtrahend	7	?	23	31	
Difference	34	26		10	2

Number42

 # 5

Addend	?	27	19	?	3
Addend	7	?	23	31	?
Sum	42	42	?	42	42

 # 6

Minuend	?	42	42	?	42
Subtrahend	7	?	23	31	?
Difference	35	27	?	11	3

Grade 2 Chapter 4 Lesson 4

Summary 4

Add or Subtract numbers in row

Number 20

6 + 14 = 14 + 6 =10+4+6=10+10= 20

20 – 14 = 20 – 10 – 4 = 10 – 4 = 6

Number 21

7 + 14 = 14 + 7 = 10 + 4 + 7 =10 + 11 = 21

21 – 14 = 21 – 11 – 3 = 10 – 3 = 7

Number 22

4 + 18 = 18 + 4 = 10 + 8 + 4 = 10 + 12 = 22

22 – 14 = 22 – 12 – 2 = 10 – 2 = 8

Number 23

9 + 14 = 14 + 9 = 10 + 4 + 9=10+13 = 23

23 – 14 = 23 – 13 – 1 = 10 – 1 = 9

Number 24

10 + 14 = 10 + 10 + 4 = 20 + 4 = 24

24 – 14 = 24 – 10 – 4 = 14 – 4 = 10

Number 25

11 + 14 = 10 + 1 + 10 + 4 = 20 + 5 = 25

25 – 14 = 25 – 10 – 4 = 15 – 4 = 11

Numbers 26

12 + 14 = 10 + 2 + 10 + 4 = 20 + 6 = 26

26 - 14 = 26 – 10 – 4 = 16 – 4 = 12

Number 27

13 + 14 = 10 + 3 + 10 + 4 = 20 + 7 = 27

27 – 14 = 27 – 10 – 4 = 17 – 4 = 13

Number 28

14 + 14 = 10 + 4 + 10 + 4 = 20 + 8 = 28

28 – 14 = 28 – 10 – 4 = 18 – 4 = 14

Number 29

15 + 14 = 10 + 5 + 10 + 4 = 20 + 9 = 29

29 – 14 = 29 – 10 – 4 = 19 – 4 = 15

Grade 2 Chapter 4 Lesson 5 Complete the tables
Number 40 Examples

a) ? + 3 = 5   5 – 3 = 2 c) 2 + ? = 5   5 – 2 = 3
b) ? - 2 = 3   3 + 2 = 5 d) 5 - ? = 3   5 – 3 = 2

1

	a	b	c	d	e
Addend		32	24		8
Addend	0		16	24	
Sum	40	40		40	40

2

Addend		31	23		7
Addend	1		17	25	
Sum	40	40		40	40

3

Addend		30	22		6
Addend	2		18	26	
Sum	40	40		40	40

4

Addend		29	21		5
Addend	3		19	27	
Sum	40	40		40	40

5

Addend		28	20		4
Addend	4		20	28	
Sum	40	40		40	40

6   d) ? + 29 = 40

				d	
Addend		27	19	?	3
Addend	5		21	29	
Sum	40	40		40	40

7   e) 2 + ? = 40

	a	b	c	d	e
Addend		26	18		2
Addend	6		22	30	?
Sum	40	40		40	40

8

Addend	?	25	17	?	1
Addend	7	?	23	31	?
Sum	40	40	?	40	40

Complete the tables

? + 3 = 5 ... + ? = 5   5 – 2 = 3

? - 2 = 3 ... - ? = 3   5 – 3 = 2

	e
Minuend	40
Subtrahend	
Difference	8

Minuend	40
Subtrahend	
Difference	7

Minuend		40
Subtrahend	26	
Difference	14	6

Minuend		40
Subtrahend	27	
Difference	13	5

Minuend		40
Subtrahend	28	
Difference	12	4

d) ? – 29 =11

	d	
Minuend	?	40
Subtrahend	29	
Difference	11	3

e) 40 - ? = 2

	d	e
Minuend		40
Subtrahend	30	?
Difference	10	2

Minuend	?	40
Subtrahend	31	?
Difference	9	1

Grade 2 Chapter 4 Lesson 6 Complete the tables
Number 41 Examples

a) ? + 3 = 5 5 – 3 = 2 c) 2 + ? = 5 5 – 2 = 3

b) ? - 2 = 3 3 + 2 = 5 d) 5 - ? = 3 5 – 3 = 2

1

	a	b	c	d	e
Addend		33	25		9
Addend	0		16	24	
Sum	41	41		41	41

2

Addend		32	24		8
Addend	1		17	25	
Sum	41	41		41	41

3

Addend		31	23		7
Addend	2		18	26	
Sum	41	41		41	41

4

Addend		30	22		6
Addend	3		19	27	
Sum	41	41		41	41

5

Addend		29	21		5
Addend	4		20	28	
Sum	41	41		41	41

6 d) ? + 29 = 41

				d	
Addend		28	20	?	4
Addend	5		21	29	
Sum	41	41		41	41

7 e) 3 + ? = 41

	a	b	c	d	e
Addend		27	19		3
Addend	6		22	30	?
Sum	41	41		41	41

8

Addend	?	26	18	?	2
Addend	7	?	23	31	?
Sum	41	41	?	41	41

Grade 2 Chapter 4 Lesson 6 Complete the tables

Number 41 Examples

a) ? + 3 = 5   5 – 3 = 2 c) 2 + ? = 5   5 – 2 = 3

b) ? - 2 = 3   3 + 2 = 5 d) 5 - ? = 3   5 – 3 = 2

9

	a	b	c	d	e
Minuend		41	41		41
Subtrahend	0		16	24	
Difference	41	33		17	9

10

Minuend		41	41		41
Subtrahend	1		17	25	
Difference	40	32		16	8

11

Minuend		41	41		41
Subtrahend	2		18	26	
Difference	39	31		15	7

12

Minuend		41	41		41
Subtrahend	3		19	27	
Difference	38	30		14	6

13

Minuend		41	41		41
Subtrahend	4		20	28	
Difference	37	29		13	5

14   d) ? – 29 = 12

				d	
Minuend		41	41	?	41
Subtrahend	5		21	29	
Difference	36	28		12	4

15   e) 41 - ? = 3

					e
Minuend		41	41		41
Subtrahend	6		22	30	?
Difference	35	27		11	3

16

Minuend	?	41	41	?	41
Subtrahend	7	?	23	31	?
Difference	34	26	?	10	2

Grade 2 Chapter 4 Lesson 7 Complete the tables
Number 42 Examples

a) ? + 3 = 5   5 – 3 = 2   c) 2 + ? = 5   5 – 2 = 3

b) ? - 2 = 3   3 + 2 = 5   d) 5 - ? = 3   5 – 3 = 2

1

	a	b	c	d	e
Addend		34	26		10
Addend	0		16	24	
Sum	42	42		42	42

2

Addend		33	25		9
Addend	1		17	25	
Sum	42	42		42	42

3

Addend		32	24		8
Addend	2		18	26	
Sum	42	42		42	42

4

Addend		31	23		7
Addend	3		19	27	
Sum	42	42		42	42

5

Addend		30	22		6
Addend	4		20	28	
Sum	42	42		42	42

6

d) ? + 29 = 42

				d	
Addend		29	21	?	5
Addend	5		21	29	
Sum	42	42		42	42

7

e) 4 + ? = 42

					e
Addend		28	20		4
Addend	6		22	30	?
Sum	42	42		42	42

8

Addend	?	27	19	?	3
Addend	7	?	23	31	?
Sum	42	42	?	42	42

Gra... hapter ... 7 Complete the tables

...les

? + 3 ... 5 – ... + ? = 5 5 – 2 = 3

? - 3 ... 3 – ... - ? = 3 5 – 3 = 2

		d	e
inu			42
ubtr		24	
iffer	2	18	10

inue			42
ubtr		25	
iffer	1	17	9

inue			42
ubtr		26	
iffer	0	16	8

inue			42
ubtr		27	
iffer	9	15	7

inue			42
ubtr		28	
iffer	8	14	6

d) ? – 29 = 13

		d	
inu		?	42
ubtr		29	
iffer	7	13	5

e) 42 - ? = 4

		d	e
inue			42
ubtr		30	?
iffer	6	12	4

inue		?	42
ubtr		31	?
iffer	5	11	3

Grade 2 Chapter 4 Lesson 8 Complete the tables
Number 43 Examples

a) ? + 3 = 5   5 – 3 = 2 c) 2 + ? = 5   5 – 2 = 3

b) ? - 2 = 3   3 + 2 = 5 d) 5 - ? = 3   5 – 3 = 2

1

	a	b	c	d	e
Addend		35	27		11
Addend	0		16	24	
Sum	43	43		43	43

2

Addend		34	26		10
Addend	1		17	25	
Sum	43	43		43	43

3

Addend		33	25		9
Addend	2		18	26	
Sum	43	43		43	43

4

Addend		32	24		8
Addend	3		19	27	
Sum	43	43		43	43

5

Addend		31	23		7
Addend	4		20	28	
Sum	43	43		43	43

6   d) ? + 29 = 43

				d	
Addend		30	22	?	6
Addend	5		21	29	
Sum	43	43		43	43

7   e) 5 + ? = 43

					e
Addend		29	21		5
Addend	6		22	30	?
Sum	43	43		43	43

8

Addend	?	28	20	?	4
Addend	7	?	23	31	?
Sum	43	43	?	43	43

Grade 2 Chapter 4 Lesson 8 Complete the tables
Number 43 Examples

a) ? + 3 = 5 5 – 3 = 2 c) 2 + ? = 5 5 – 2 = 3

b) ? - 2 = 3 3 + 2 = 5 d) 5 - ? = 3 5 – 3 = 2

9

	a	b	c	d	e
Minuend		43	43		43
Subtrahend	0		16	24	
Difference	43	35		19	11

10

Minuend		43	43		43
Subtrahend	1		17	25	
Difference	42	34		18	10

11

Minuend		43	43		43
Subtrahend	2		18	26	
Difference	41	33		17	9

12

Minuend		43	43		43
Subtrahend	3		19	27	
Difference	40	32		16	8

13

Minuend		43	43		43
Subtrahend	4		20	28	
Difference	39	31		15	7

14 d) ? – 29 = 14

				d	
Minuend		43	43	?	43
Subtrahend	5		21	29	
Difference	38	30		14	6

15 e) 43 - ? = 5

					e
Minuend		43	43		43
Subtrahend	6		22	30	?
Difference	37	29		13	5

16

Minuend	?	43	43	?	43
Subtrahend	7	?	23	31	?
Difference	36	28	?	12	4

Grade 2 Chapter 4 Lesson 9 Complete the tables
Number 44 Examples

a) ? + 3 = 5   5 – 3 = 2 c) 2 + ? = 5   5 – 2 = 3

b) ? - 2 = 3   3 + 2 = 5 d) 5 - ? = 3   5 – 3 = 2

1

	a	b	c	d	e
Addend		36	28		12
Addend	0		16	24	
Sum	44	44		44	44

2

Addend		35	27		11
Addend	1		17	25	
Sum	44	44		44	44

3

Addend		34	26		10
Addend	2		18	26	
Sum	44	44		44	44

4

Addend		33	25		9
Addend	3		19	27	
Sum	44	44		44	44

5

Addend		32	24		8
Addend	4		20	28	
Sum	44	44		44	44

6   d   d) ? + 29 = 44

				d	
Addend		31	23	?	7
Addend	5		21	29	
Sum	44	44		44	44

7   e   e) 6 + ? = 44

					e
Addend		30	22		6
Addend	6		22	30	?
Sum	44	44		44	44

8

Addend	?	29	21	?	5
Addend	7	?	23	31	?
Sum	44	44	?	44	44

…ide 2… …Complete the tables

…les

…? + … …+ ? = 5 5 – 2 = 3

…? - … …- ? = 3 5 – 3 = 2

		d	e
…inu…			44
…btr…		24	
…iffer…	…4	20	12

…inu…			44
…ubtr…		25	
…iffer…	3	19	11

…inu…			44
…ubtr…		26	
…iffer…	2	18	10

…inu…			44
…ubtr…		27	
…iffer…	1	17	9

…inu…			44
…ubtr…		28	
…iffer…	0	16	8

		d	
…inu…		?	44
…btr…		29	
…iffer…	…9	15	7

d) ? – 29 = 15

			e
…inu…			44
…ubtr…		30	?
…iffer…	8	14	6

e) 44 - ? = 6

…inu…		?	44
…ubtr…		31	?
…iffer…	7	13	5

Grade 2 Chapter 4 Lesson 10 Complete the tables
Number 45 Examples

a) ? + 3 = 5   5 – 3 = 2 c) 2 + ? = 5   5 – 2 = 3

b) ? - 2 = 3   3 + 2 = 5 d) 5 - ? = 3   5 – 3 = 2

1

	a	b	c	d	e
Addend		37	29		13
Addend	0		16	24	
Sum	45	45		45	45

2

Addend		36	28		12
Addend	1		17	25	
Sum	45	45		45	45

3

Addend		35	27		11
Addend	2		18	26	
Sum	45	45		45	45

4

Addend		34	26		10
Addend	3		19	27	
Sum	45	45		45	45

5

Addend		33	25		9
Addend	4		20	28	
Sum	45	45		45	45

6   d) ? + 29 = 45

				d	
Addend		32	24	?	8
Addend	5		21	29	
Sum	45	45		45	45

7   e) 7 + ? = 45

					e
Addend		31	23		7
Addend	6		22	30	?
Sum	45	45		45	45

8

Addend	?	30	22	?	6
Addend	7	?	23	31	?
Sum	45	45	?	45	45

Grade 2 Chapter 4 Lesson 10 Complete the tables
Number 45 Examples

a) ? + 3 = 5   5 – 3 = 2 c) 2 + ? = 5   5 – 2 = 3
b) ? - 2 = 3   3 + 2 = 5 d) 5 - ? = 3   5 – 3 = 2

9

	a	b	c	d	e
Minuend		45	45		45
Subtrahend	0		16	24	
Difference	45	37		21	13

10

Minuend		45	45		45
Subtrahend	1		17	25	
Difference	44	36		20	12

11

Minuend		45	45		45
Subtrahend	2		18	26	
Difference	43	35		19	11

12

Minuend		45	45		45
Subtrahend	3		19	27	
Difference	42	34		18	10

13

Minuend		45	45		45
Subtrahend	4		20	28	
Difference	41	33		17	9

14 d) ? – 29 = 16

				d	
Minuend		45	45	?	45
Subtrahend	5		21	29	
Difference	40	32		16	8

15 e) 45 - ? = 7

					e
Minuend		45	45		45
Subtrahend	6		22	30	?
Difference	39	31		15	7

16

Minuend	?	45	45	?	45
Subtrahend	7	?	23	31	?
Difference	38	30	?	14	6

Grade 2 Chapter 4 Lesson 11 Complete the tables
Number 46 Examples

a) ? + 3 = 5   5 – 3 = 2 c) 2 + ? = 5   5 – 2 = 3

b) ? - 2 = 3   3 + 2 = 5 d) 5 - ? = 3   5 – 3 = 2

1

	a	b	c	d	e
Addend		38	30		14
Addend	0		16	24	
Sum	46	46		46	46

2

Addend		37	29		13
Addend	1		17	25	
Sum	46	46		46	46

3

Addend		36	28		12
Addend	2		18	26	
Sum	46	46		46	46

4

Addend		35	27		11
Addend	3		19	27	
Sum	46	46		46	46

5

Addend		34	26		10
Addend	4		20	28	
Sum	46	46		46	46

6   d) ? + 29 = 46

				d	
Addend		33	25	?	9
Addend	5		21	29	
Sum	46	46		46	46

7   e) 8 + ? = 46

					e
Addend		32	24		8
Addend	6		22	30	?
Sum	46	46		46	46

8

Addend	?	31	23	?	7
Addend	7	?	23	31	?
Sum	46	46	?	46	46

Grade ... 11 Complete the tables

...les

? + ... + ? = 5   5 – 2 = 3

? - ... - ? = 3   5 – 3 = 2

	d	e
inu...		46
ubtra...	24	
iffer...	22	14

inu...		46
ubtra...	25	
iffer...	21	13

inu...		46
ubtra...	26	
iffer...	20	12

inu...		46
ubtra...	27	
iffer...	19	11

inu...		46
ubtra...	28	
iffer...	18	10

d) ? – 29 = 17

	d	
inu...	?	46
ubtra...	29	
iffer...	17	9

e) 46 - ? = 8

		e
inu...		46
ubtra...	30	?
iffer...	16	8

inu...	?	46
ubtra...	31	?
iffer...	15	7

Grade 2 Chapter 4 Lesson 12 Complete the tables
Number 47 Examples

a) ? + 3 = 5   5 – 3 = 2 c) 2 + ? = 5   5 – 2 = 3

b) ? - 2 = 3   3 + 2 = 5 d) 5 - ? = 3   5 – 3 = 2

1

	a	b	c	d	e
Addend		39	31		15
Addend	0		16	24	
Sum	47	47		47	47

2

Addend		38	30		14
Addend	1		17	25	
Sum	47	47		47	47

3

Addend		37	29		13
Addend	2		18	26	
Sum	47	47		47	47

4

Addend		36	28		12
Addend	3		19	27	
Sum	47	47		47	47

5

Addend		35	27		11
Addend	4		20	28	
Sum	47	47		47	47

6

				d	
Addend		34	26	?	10
Addend	5		21	29	
Sum	47	47		47	47

d) ? + 29 = 47

7

					e
Addend		33	25		9
Addend	6		22	30	?
Sum	47	47		47	47

e) 9 + ? = 47

8

Addend	?	32	24	?	8
Addend	7	?	23	31	?
Sum	47	47	?	47	47

Grade 2 Chapter 4 Lesson 12 Complete the tables
Number 47 Examples

a) ? + 3 = 5   5 – 3 = 2 c) 2 + ? = 5   5 – 2 = 3
b) ? - 2 = 3   3 + 2 = 5 d) 5 - ? = 3   5 – 3 = 2

9

	a	b	c	d	e
Minuend		47	47		47
Subtrahend	0		16	24	
Difference	47	39		23	15

10

Minuend		47	47		47
Subtrahend	1		17	25	
Difference	46	38		22	14

11

Minuend		47	47		47
Subtrahend	2		18	26	
Difference	45	37		21	13

12

Minuend		47	47		47
Subtrahend	3		19	27	
Difference	44	36		20	12

13

Minuend		47	47		47
Subtrahend	4		20	28	
Difference	43	35		19	11

14   d) ? – 29 = 18

				d	
Minuend		47	47	?	47
Subtrahend	5		21	29	
Difference	42	34		18	10

15   e) 47 – ? = 9

	a	b	c	d	e
Minuend		47	47		47
Subtrahend	6		22	30	?
Difference	41	33		17	9

16

Minuend	?	47	47	?	47
Subtrahend	7	?	23	31	?
Difference	40	32	?	16	8

Grade 2 Chapter 4 Lesson 13 Complete the tables
Number 48 Examples

a) ? + 3 = 5   5 – 3 = 2   c) 2 + ? = 5   5 – 2 = 3

b) ? - 2 = 3   3 + 2 = 5   d) 5 - ? = 3   5 – 3 = 2

1

	a	b	c	d	e
Addend		40	32		16
Addend	0		16	24	
Sum	48	48		48	48

2

Addend		39	31		15
Addend	1		17	25	
Sum	48	48		48	48

3

Addend		38	30		14
Addend	2		18	26	
Sum	48	48		48	48

4

Addend		37	29		13
Addend	3		19	27	
Sum	48	48		48	48

5

Addend		36	28		12
Addend	4		20	28	
Sum	48	48		48	48

6   d) ? + 29 = 48

				d	
Addend		35	27	?	11
Addend	5		21	29	
Sum	48	48		48	48

7   e) 10 + ? = 48

					e
Addend		34	26		10
Addend	6		22	30	?
Sum	48	48		48	48

8

Addend	?	33	25	?	9
Addend	7	?	23	31	?
Sum	48	48	?	48	48

13 Complete the tables

les

+ ? = 5   5 – 2 = 3

- ? = 3   5 – 3 = 2

	d	e
inu		48
ibtr	24	
iffe	24	16

inu		48
ubtr	25	
iffe	23	15

inu		48
ubtr	26	
iffe	22	14

inu		48
ubtr	27	
iffe	21	13

inu		48
ubtr	28	
iffe	20	12

d) ? - 29 = 19

	d	
inu	?	48
ibtr	29	
iffe	19	11

e) 48 – ? = 10

		e
inu		48
ubtr	30	?
iffe	18	10

inu	?	48
ubtr	31	?
iffe	17	9

Grade 2 Chapter 4 Lesson 14 Complete the tables

Number 49 Examples

a) ? + 3 = 5   5 – 3 = 2 c) 2 + ? = 5   5 – 2 = 3

b) ? - 2 = 3   3 + 2 = 5 d) 5 - ? = 3   5 – 3 = 2

 # 1

	a	b	c	d	e
Addend		41	33		17
Addend	0		16	24	
Sum	49	49		49	49

 # 2

Addend		40	32		16
Addend	1		17	25	
Sum	49	49		49	49

 # 3

Addend		39	31		15
Addend	2		18	26	
Sum	49	49		49	49

 # 4

Addend		38	30		14
Addend	3		19	27	
Sum	49	49		49	49

 # 5

Addend		37	29		13
Addend	4		20	28	
Sum	49	49		49	49

 # 6

d) ? + 29 = 49

				d	
Addend		36	28	?	12
Addend	5		21	29	
Sum	49	49		49	49

 # 7

e) 11 + ? = 49

					e
Addend		35	27		11
Addend	6		22	30	?
Sum	49	49		49	49

 # 8

Addend	?	34	26	?	10
Addend	7	?	23	31	?
Sum	49	49	?	49	49

Grade 2 Chapter 4 Lesson 14 Complete the tables
Number 49 Examples

a) ? + 3 = 5   5 – 3 = 2 c) 2 + ? = 5   5 – 2 = 3
b) ? - 2 = 3   3 + 2 = 5 d) 5 - ? = 3   5 – 3 = 2

9

	a	b	c	d	e
Minuend		49	49		49
Subtrahend	0		16	24	
Difference	49	41		25	17

10

Minuend		49	49		49
Subtrahend	1		17	25	
Difference	48	40		24	16

11

Minuend		49	49		49
Subtrahend	2		18	26	
Difference	47	39		23	15

12

Minuend		49	49		49
Subtrahend	3		19	27	
Difference	46	38		22	14

13

Minuend		49	49		49
Subtrahend	4		20	28	
Difference	45	37		21	13

14

				d	
Minuend		49	49	?	49
Subtrahend	5		21	29	
Difference	44	36		20	12

d) ? – 29 = 20

15

					e
Minuend		49	49		49
Subtrahend	6		22	30	?
Difference	43	35		19	11

e) 49 – ? = 11

16

Minuend	?	49	49	?	49
Subtrahend	7	?	23	31	?
Difference	42	34	?	18	10

Grade 2 Chapter 4 Lesson 15
Advanced Problems
Numbers from 40 to 49
Test 4 A
Complete the tables

# 1	a	b	c	d	e
Addend		32	24		8
Addend	0		16	24	
Sum	40	40		40	40

2

Addend		32	24		8
Addend	1		17	25	
Sum	41	41		41	41

3

Addend	?	32	24	?	8
Addend	2	?	18	26	?
Sum	42	42	?	42	42

# 4	a	b	c	d	e
Minuend	?	42	42	?	42
Subtrahend	2	?	18	26	?
Difference	40	32	?	16	8

5

Minuend		43	43		43
Subtrahend	3		19	27	
Difference	40	32		16	8

6

Minuend		44	44		44
Subtrahend	4		20	28	
Difference	40	32		16	8

Lesson 15

to 49

 # 1

	d	e
end		8
end	27	
[illegible]	43	43

 # 2

	d	e
end		8
end	28	
[illegible]	44	44

 # 3

	d	e
end	?	8
end	29	?
[illegible]	45	45

 # 4

	d	e
end		45
ahen	29	
rence	16	8

 # 5

	d	e
end		46
rahen	30	
rence	16	8

 # 6

	d	e
end		47
rahen	31	
rence	16	8

Grade 2 Chapter 4 Lesson 16
Review 2 A (Basic Problems)
Numbers # 1 - # 20 odd or # 1, # 3, # 5, # 7, # 9, # 11, # 13, # 15, # 17, # 19
Add or Subtract Numbers in Row

1 31 + 39

2 39 + 32

3 33 + 39

4 39 + 34

5 35 + 39

6 70 – 31

7 71 – 39

8 72 – 33

9 73 – 39

10 74 – 35

Add or Subtract Numbers in Column

# 11	# 12	# 13	# 14	# 15
36	39	38	39	40
+39	+37	+ 39	+39	+ 39

# 16	# 17	# 18	# 19	# 20
75	76	77	78	79
-39	- 37	-39	-49	- 29

Grade 2 Chapter 4 Lesson 16
Review 2 B (Basic Problems)
Put right number instead question mark
Numbers # 1 - # 20 odd or # 1, # 3, # 5, # 7, # 9, # 11, # 13, # 15, # 17, # 19

1 41 + 39

2 39 + 42

3 43 + 39

4 39 + 44

5 35 + 49

6 80 – 39

7 81 - 32

8 82 – 39

9 83 - 34

10 84 – 39

Add or Subtract Numbers in Column

# 11	# 12	# 13	# 14	# 15
36	49	48	39	50
+49	+37	+ 39	+49	+ 39

# 16	# 17	# 18	# 19	# 20
85	86	87	88	89
-39	-37	-39	-49	- 29

Grade 2 Chapter 4 Lesson 17
Final Exam 2 A (Basic Problems)
Numbers # 1 - # 20 odd or # 1, # 3, # 5, # 7, # 9, # 11, # 13, # 15, # 17, # 19
Add or Subtract Numbers in Row

1 11 + 39

2 29 + 22

3 23 + 29

4 29 + 24

5 15 + 39

6 50 – 31

7 51 – 29

8 52 – 13

9 53 – 29

10 54 – 25

Add or Subtract Numbers in Column

# 11 29 +26	# 12 29 +27	# 13 28 + 29	# 14 19 +39	# 15 40 + 19
# 16 55 -28	# 17 56 -38	# 18 57 -18	# 19 58 -39	# 20 59 - 19

Gr[illegible] Chap[illegible] [illegible]son 17
Fin[illegible] [illegible] B[illegible] [illegible]oblems)
Numbers # 1 - [illegible] # 1, [illegible] # [illegible], # 13, # 15, # 17, # 19
Add [illegible] [illegible]act[illegible] in Row

1 29 + 31
2 32 + 29

3 39 + 23
4 34 + 29

5 29 + 35
6 60 – 31

7 61 - 22
8 62 – 13

9 63 - 24
10 64 – 29

Add [illegible] [illegible]act [illegible] in C[illegible]

# 11 29 +36	[illegible]	[illegible]	[illegible]	# 15 40 + 29
# 16 65 -26	[illegible]	[illegible]	[illegible]	# 20 69 - 29

 # 1 - # 20 odd o[illegible] 9, [illegible]
Grading Scale: [illegible] [illegible]nts, [illegible] [illegible], D=7 points, F=0-6 points

Final Exam 2 B[illegible] [illegible]blem[illegible]

# 1	60	# [illegible]	[illegible]	# 9	[illegible]	# 13	[illegible]	# 17	29
# 2	61	# [illegible]	[illegible]	# 10	[illegible]	# 14	[illegible]	# 18	38
# 3	62	# [illegible]	[illegible]	# 11	[illegible]	# 15	[illegible]	# 19	39
# 4	63	# [illegible]	[illegible]	# 12	[illegible]	# 16	[illegible]	# 20	40

Grade 2 Answers Chapter 4

Lesson 1	a	b	c	d	Summary 1	a	b	c	d
# 1	28	1	30	13	# 11	35	29	5	31
# 2	2	29	30	16	# 12	35	6	30	4
# 3	28	2	30	0	# 13	36	29	6	31
# 4	28	2	30	0	# 14	36	29	6	31
# 5	4	29	2	31	# 15	37	29	7	31
# 6	28	3	30	1	# 16	37	29	7	31
# 7	5	29	3	31	# 17	38	29	8	31
# 8	28	4	30	2	# 18	38	29	8	31
# 9	6	29	4	31	# 19	39	29	9	31
# 10	28	5	30	3	# 20	39	29	9	31

Lesson 2 Summary 2 Review 1 A

	a	b	c	d	e		a	b	c	d	e
# 1	3	6	7	8	6	# 3	9	7	17	10	8
# 2	10	9	12	4	14	# 4	15	8	9	18	10

Review 1 B

	a	b	c	d	e		a	b	c	d	e
# 1	2	7	8	9	5	# 3	7	8	17	9	10
# 2	10	8	12	5	14	# 4	15	7	8	18	11

Lesson 3 Summary 3

	a	b	c	d	e		a	b	c	d	e
# 1	33	15	40	9	39	# 4	41	15	18	41	39
# 2	40	15	17	40	39	# 5	35	15	42	11	39
# 3	34	15	41	10	39	# 6	42	15	19	42	39

Lesson 5	a b c d e	Number 40	a b c d e
# 1	40, 8, 40, 16, 32	# 9	40, 8, 24, 40, 32
# 2	39, 9, 40, 15, 33	# 10	40, 9, 23, 40, 33
# 3	38, 10, 40, 14, 34	# 11	40, 10, 22, 40, 34
# 4	37, 11, 40, 13, 35	# 12	40, 11, 21, 40, 35
# 5	36, 12, 40, 12, 36	# 13	40, 12, 20, 40, 36
# 6	35, 13, 40, 11, 37	# 14	40, 13, 19, 40, 37
# 7	34, 14, 40, 10, 38	# 15	40, 14, 18, 40, 38
# 8	33, 15, 40, 9, 39	# 16	40, 15, 17, 40, 39

Grade 2 Answers Chapter 4

Lesson 6

1 41, 8, 41, 17, 32
2 40, 9, 41, 16, 33
3 39, 10, 41, 15, 34
4 38, 11, 41, 14, 35
5 37, 12, 41, 13, 36
6 36, 13, 41, 12, 37
7 35, 14, 41, 11, 38
8 34, 15, 41, 10, 39

Lesson 7

1 42, 8, 42, 18, 32
2 41, 9, 42, 17, 33
3 40, 10, 42, 16, 34
4 39, 11, 42, 15, 35
5 38, 12, 42, 14, 36
6 37, 13, 42, 13, 37
7 36, 14, 42, 12, 38
8 35, 15, 42, 11, 39

Lesson 8

1 43, 8, 43, 19, 32
2 42, 9, 43, 18, 33
3 41, 10, 43, 17, 34
4 40, 11, 43, 16, 35
5 39, 12, 43, 15, 36
6 38, 13, 43, 14, 37
7 37, 14, 43, 13, 38
8 36, 15, 43, 12, 39

Lesson 9

a b c d e
1 44, 8, 44, 20, 32
2 43, 9, 44, 19, 33
3 42, 10, 44, 18, 34
4 41, 11, 44, 17, 35
5 40, 12, 44, 16, 36
6 39, 13, 44, 15, 37
7 38, 14, 44, 14, 38
8 37, 15, 44, 13, 39

Number 41

9 41, 8, 25, 41, 32
10 41, 9, 24, 41, 33
11 41, 10, 23, 41, 34
12 41, 11, 22, 41, 35
13 41, 12, 21, 41, 36
14 41, 13, 20, 41, 37
15 41, 14, 19, 41, 38
16 41, 15, 18, 41, 39

Number 42

9 42, 8, 26, 42, 32
10 42, 9, 25, 42, 33
11 42, 10, 24, 42, 34
12 42, 11, 23, 42, 35
13 42, 12, 22, 42, 36
14 42, 13, 21, 42, 37
15 42, 14, 20, 42, 38
16 42, 15, 19, 42, 39

Number 43

9 43, 8, 27, 43, 32
10 43, 9, 26, 43, 33
11 43, 10, 25, 43, 34
12 43, 11, 24, 43, 35
13 43, 12, 23, 43, 36
14 43, 13, 22, 43, 37
15 43, 14, 21, 43, 38
16 43, 15, 20, 43, 39

Number 44

a b c d e
9 44, 8, 28, 44, 32
10 44, 9, 27, 44, 33
11 44, 10, 26, 44, 34
12 44, 11, 25, 44, 35
13 44, 12, 24, 44, 36
14 44, 13, 23, 44, 37
15 44, 14, 22, 44, 38
16 44, 15, 21, 44, 39

Grade 2 Answers Chapter 4

Lesson 10

# 1	45, 8, 45, 21, 32
# 2	44, 9, 45, 20, 33
# 3	43, 10, 45, 19, 34
# 4	42, 11, 45, 18, 35
# 5	41, 12, 45, 17, 36
# 6	40, 13, 45, 16, 37
# 7	39, 14, 45, 15, 38
# 8	38, 15, 45, 14, 39

Lesson 11

# 1	46, 8, 46, 22, 32
# 2	45, 9, 46, 21, 33
# 3	44, 10, 46, 20, 34
# 4	43, 11, 46, 19, 35
# 5	42, 12, 46, 18, 36
# 6	41, 13, 46, 17, 37
# 7	40, 14, 46, 16, 38
# 8	39, 15, 46, 15, 39

Lesson 12

# 1	47, 8, 47, 23, 32
# 2	46, 9, 47, 22, 33
# 3	45, 10, 47, 21, 34
# 4	44, 11, 47, 20, 35
# 5	43, 12, 47, 19, 36
# 6	42, 13, 47, 18, 37
# 7	41, 14, 47, 17, 38
# 8	40, 15, 47, 16, 39

Lesson 13

	a	b	c	d	e
# 1	48,	8,	48,	24,	32
# 2	47,	9,	48,	23,	33
# 3	46,	10,	48,	22,	34
# 4	45,	11,	48,	21,	35
# 5	44,	12,	48,	20,	36
# 6	43,	13,	48,	19,	37
# 7	42,	14,	48,	18,	38
# 8	41,	15,	48,	17,	39

Number 45

# 9	45, 8, 29, 45, 32
# 10	45, 9, 28, 45, 33
# 11	45, 10, 27, 45, 34
# 12	45, 11, 26, 45, 35
# 13	45, 12, 25, 45, 36
# 14	45, 13, 24, 45, 37
# 15	45, 14, 23, 45, 38
# 16	45, 15, 22, 45, 39

Number 46

# 9	46, 8, 30, 46, 32
# 10	46, 9, 29, 46, 33
# 11	46, 10, 28, 46, 34
# 12	46, 11, 27, 46, 35
# 13	46, 12, 26, 46, 36
# 14	46, 13, 25, 46, 37
# 15	46, 14, 24, 46, 38
# 16	46, 15, 23, 46, 39

Number 47

# 9	47, 8, 31, 47, 32
# 10	47, 9, 30, 47, 33
# 11	47, 10, 29, 47, 34
# 12	47, 11, 28, 47, 35
# 13	47, 12, 27, 47, 36
# 14	47, 13, 26, 47, 37
# 15	47, 14, 25, 47, 38
# 16	47, 15, 24, 47, 39

Number 48

	a	b	c	d	e
# 9	48,	8,	32,	48,	32
# 10	48,	9,	31,	48,	33
# 11	48,	10,	30,	48,	34
# 12	48,	11,	29,	48,	35
# 13	48,	12,	28,	48,	36
# 14	48,	13,	27,	48,	37
# 15	48,	14,	26,	48,	38
# 16	48,	15,	25,	48,	39

Lesson 14 [illegible] 49

1 49, 8, 49, [illegible] 8, [illegible] 2
2 48, 9, 49, [illegible] 9, [illegible] 3
3 47, 10, 49, [illegible] 10, [illegible] 4
4 46, 11, 49, [illegible] 11, [illegible] 5
5 45, 12, 49, [illegible] 12, [illegible] 6
6 44, 13, 49, [illegible] 13, [illegible] 7
7 43, 14, 49, [illegible] 14, [illegible] 8
8 42, 15, 49, [illegible] 15, [illegible] 9

Lesson 15

Test 4 A [illegible]

a b c [illegible] b [illegible] e
1 40 8 40 [illegible] 11 [illegible] 35
2 40 9 41 [illegible] 36
3 40 10 42 [illegible] 13 [illegible] 37
4 42 10 24 [illegible] 13 [illegible] 37
5 43 11 24 [illegible] 14 [illegible] 38
6 44 12 24 [illegible] 15 [illegible] 39

Lesson 16 Revi[illegible]**sic Pr**[illegible]

1 70 # 5 [illegible] 9 [illegible] 3 [illegible] 7 39
2 71 # 6 [illegible] 10 [illegible] 4 [illegible] 8 38
3 72 # 7 [illegible] 11 [illegible] 5 [illegible] 9 29
4 73 # 8 [illegible] 12 [illegible] 6 [illegible] 0 50

Review 2 B (Ba[illegible]**ms)**

1 80 # 5 [illegible] 9 [illegible] 3 [illegible] 7 49
2 81 # 6 [illegible] 10 [illegible] 4 [illegible] 8 48
3 82 # 7 [illegible] 11 [illegible] 5 [illegible] 9 39
4 83 # 8 [illegible] 12 [illegible] 6 [illegible] 60

Lesson 17 Final [illegible] **(Ba**[illegible]**ems)**

1 50 # 5 [illegible] 9 [illegible] 3 [illegible] 7 18
2 51 # 6 [illegible] 10 [illegible] 4 [illegible] 8 39
3 52 # 7 [illegible] 11 [illegible] 5 [illegible] 9 19
4 53 # 8 [illegible] 12 [illegible] 6 [illegible] 40

Final Exam 2 B [illegible]**blem**[illegible]

1 60 # [illegible] # 9 [illegible] 3 [illegible] 7 29
2 61 # [illegible] # 10 [illegible] 4 [illegible] 8 38
3 62 # [illegible] # 11 [illegible] 5 [illegible] 9 39
4 63 # [illegible] # 12 [illegible] 6 [illegible] 0 40

Students!! Attention!!

Adults, help please students 1 – 4 grades to read and understand advice.

When I was in 5th grade I did not like math. I am not always understood math. I was 11 years girl. Next 6th grade I got a very good teacher. She explains very clearly and I understood her very well. I was worked very hard and I was not afraid to ask questions. Final Grade in Algebra in 6th grade I got an A. I said in class that I want to be a teacher of math. But some students said that I am not enough smart to be a math teacher. I said I will be work hard with math. I did it, I got a Master's degree in math and education, I love my students, I was teaching many years, in High School, part-time college math instructor, full-time university math lecturer.

Some people do a simple job, eat, sleep, go to work. People with a bad brain have not so a good job, not so good salary. If you look at TV how someone's dance, you will be not changing your brain. But if you take a dancing class this helps your body and brain ability. The same if you want to use the calculator, but the calculator not always with you. Much better if you learn how you need to do any operations with numbers without a calculator. If you read this math book you will know what you need to do. This teaches you how to improve your memory, attention, logic thinking, how to read your math books every day, your brain works better. You can work 50 minutes with math, 10 minutes break maybe dance. Do not afraid to do mistakes, fixed and move to the next step. Don't listen to anyone, choose what do you like more.

Students need to know if you work with the brain every day, your brain works better. You need a good brain in any job, in any class. When you finish High School, you need to get any profession, you need to go to college and get a good job and a better salary. You need to learn any subject, its work for the brain. Read the book, do the exercises yourself, try your best, and then check your answer at the end of the chapter. Be honest with yourself. You need to believe in yourself. The brain is like muscles. People with good brain lives longer. The professors at my university were over 80 years old, wise, and optimistic. It was a great pleasure to talk to them.

Made in the USA
Monee, IL
05 November 2026

67380397R00063